Flows and Frictions in Trans-Himalayan Spaces

Flows and Frictions in Trans-Himalayan Spaces

Histories of Networking and Border Crossing

Edited by
Gunnel Cederlöf and
Willem van Schendel

Routledge
Taylor & Francis Group
LONDON AND NEW YORK

First published in 2022 by Amsterdam University Press Ltd.

Published 2025 by Routledge
4 Park Square, Milton Park, Abingdon, Oxon OX14 4RN
605 Third Avenue, New York, NY 10158

Routledge is an imprint of the Taylor & Francis Group, an informa business

ISBN: 9789463724371 (hbk)
ISBN: 9781041179511 (pbk)
ISBN: 9781003695769 (ebk)
NUR 906

Cover photo: Navigating a landslide on a mountain road, Mizoram (India), 1940s-1950s
Courtesy the Lalhruaitluanga Ralte collection (Aizawl, Mizoram, India); photographer unidentified

Cover design: Coördesign, Leiden

DOI 10.5117/9789463724371

For Product Safety Concerns and Information please contact our EU representative:
GPSR@taylorandfrancis.com
Taylor & Francis Verlag GmbH, Kaufingerstraße 24, 80331 München, Germany

Table of Contents

Acknowledgements 9

1 Flows and Frictions in Trans-Himalayan Spaces 11
An Introduction
Willem van Schendel and Gunnel Cederlöf

Prologue

2 Spatial History in Southern Asia 29
Mobility, Territoriality, and Religion
David Ludden

A Long View

3 The Road Towards *All under Heaven* Cosmology 53
The Bazi Basin Society in West Yunnan
Jianxiong Ma

4 Tracking Routes 77
Imperial Competition in the Late-nineteenth Century Burma-China Borderlands
Gunnel Cederlöf

5 'Circulations' along the Indo-Burma Borderlands 105
Networks of Trade, Religion, and Identity
Joy L. K. Pachuau

6 Flows and Fairs 137
The Eastern Himalayas and the British Empire
Arupjyoti Saikia

Mobilities Today

7 How to Interpret a Lynching? 167
Immigrant Flows, Ethnic Anxiety, and Sovereignty in Nagaland, Northeast India
Jelle J. P. Wouters

8 Frictions and Opacities in the Myanmar-China Jade Trade 203
Henrik Kloppenborg Møller

9 Multiple Identities of Young Sittwe Muslims and Becoming Rohingya 231
Tharaphi Than and Htoo May

Bibliography 255

Index 283

List of Maps and Figures

Map 0.1	The India-China Corridor	8
Map 3.1	The Shan-Dai region on the frontiers of Burma and China	61
Map 4.1	Communication routes between Bhamo and Momein, and the four zones, late 19th century	88
Map 6.1	Market towns connecting central and north Assam, eastern Tibet, Sichuan in China and Kachin, 19th and early 20th century	138
Map 8.1	Jade flows across Myanmar and China	213

All maps are drawn by Laurie Whiddon, Map Illustrations.

Figure 1.1	Photograph captioned 'Group of Wild Chins'	15
Figure 5.1	Items of Khasi jewellery in silver	113
Figure 5.2	Memorial stone (*lungdawh*) of Lalpuithanga, 18th century	115
Figure 5.3	Rock engravings at Mualcheng, Mizoram	127
Figure 9.1	Different ID cards in Myanmar	237
Figure 9.2	U Ba Tun, a Rohingya broadcaster, presenting Rohingya programmes at the Burmese Broadcasting Station	239

Map 0.1 The India-China Corridor

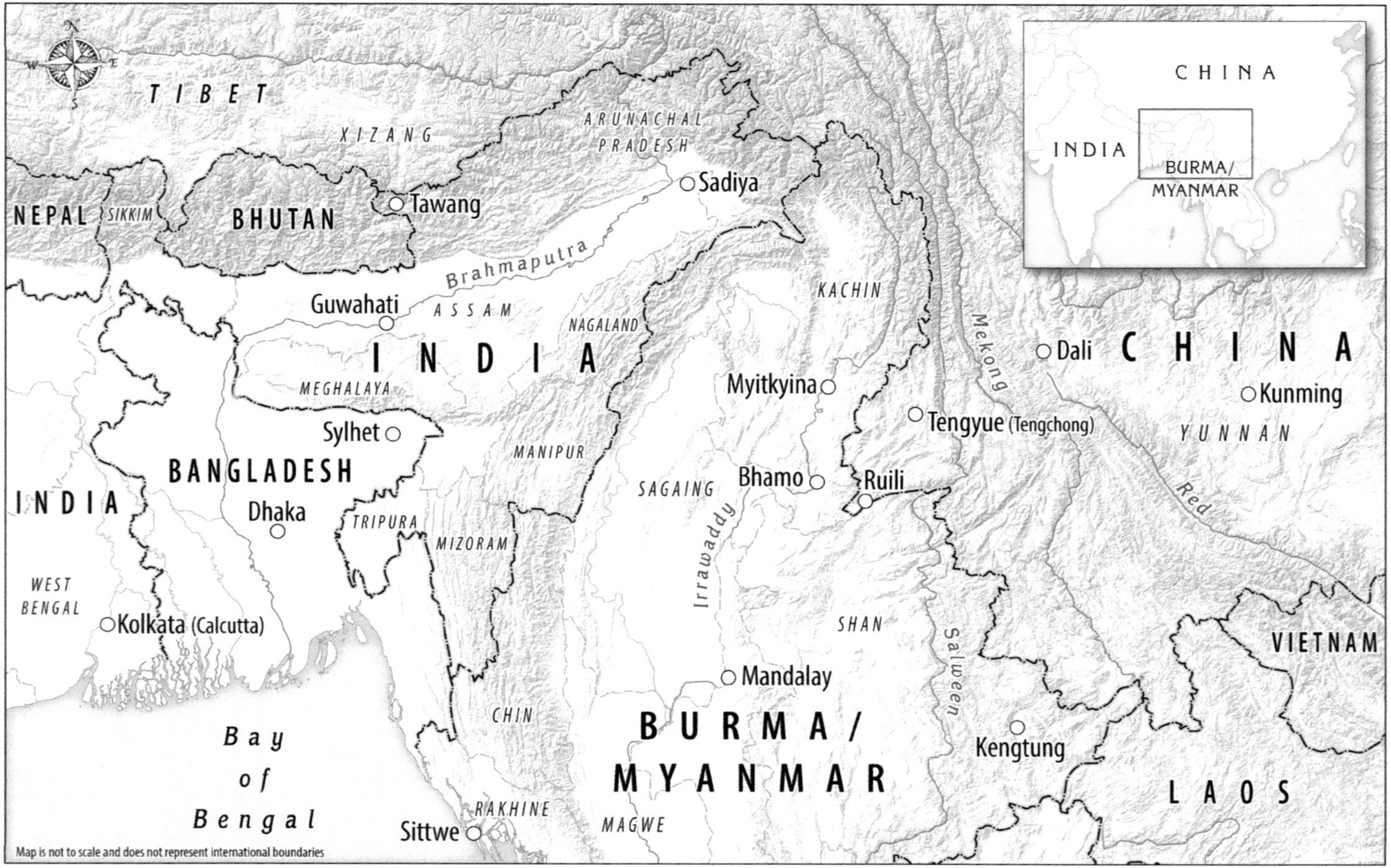

Drawn by Laurie Whiddon

Acknowledgements

The contributions to this volume have grown out of 'The India-China Corridor' research project and its final meeting with a large research network, 'Flows and Frictions in Trans-Himalayan Spaces,' in 2019. This was the last of four events at which researchers presented and discussed ongoing work in Northeast India, Myanmar, Yunnan, Tibet, Bangladesh, and Southeast Asia. We are much indebted to our project's research team members Mandy Sadan, Arupjyoti Saikia, and Dan Smyer Yü for their knowledge, inspiration, collegiality, and friendship over the years. We have benefitted greatly from discussions and feedback from the participants in all large and small network meetings, and we would like to extend our gratitude to the Indian Institute of Technology Guwahati, Yunnan Minzu University, and Kachinland Research Centre for their hospitality. We are also grateful to the Swedish Research Council that funded the project, and the Royal Swedish Academy of Letters, History and Antiquities; Riksbankens Jubileumsfond; and Linnaeus University Center for Concurrences in Colonial and Postcolonial Studies that supported the meeting in 2019. We are especially grateful to the Academy for providing administrative assistance and a venue for the event. The volume has benefitted significantly from careful copyediting and maps. Wendy Davies has added clarity to the text through her excellent copyediting, Rikard Ehnsiö edited the references, and Laurie Whiddon has drawn beautiful maps of a complex and complicated past.

1 Flows and Frictions in Trans-Himalayan Spaces

An Introduction

Willem van Schendel and Gunnel Cederlöf

Abstract
This volume presents a conversation between historians and anthropologists who work at the crosscurrents of Asian borderland studies and Trans-Himalayan studies. We focus on the evolving relationships of time, space, and place, while combining the ethnographically historical and the historically ethnographic. In line with many recent studies, this volume challenges the conventional foregrounding of nation-states (a short-lived phenomenon in the longer view) without, however, resorting to the fantasy of disembodied flows. Contemporary events require long-term perspectives since past mobilities underlie many of today's complex conflicts. Anchored in the region's stunning landscapes (but not determined by them), flows are negotiated in webs of human interaction (but not defined by them).

Keywords: Trans-Himalayan spaces, India-China corridor, flow and friction, spatial history, Myanmar, borderlands

This book deals with a vast region of rugged terrain, high mountains, monsoon-fed rivers and marshy lowlands. It is a landscape that may look like a formidable natural obstacle to mobility, but the contributions to this book show otherwise. They explore the abundance of movement, connections and flows that have given this space, straddling the eastern Himalayas, its historical shape and its contemporary dynamism.

It is not only the forbidding landscape itself that seemingly creates obstacles to movement: the region is also crisscrossed by state borders that

Cederlöf, Gunnel, and Willem van Schendel (eds), *Flows and Frictions in Trans-Himalayan Spaces: Histories of Networking and Border Crossing.* Taylor & Francis Group 2022
DOI: 10.5117/9789463724371_CH01

have hardened and become more militarised during the past century. Today, these borders separate the territories of India, Bhutan, Burma (Myanmar), Bangladesh, China, and Thailand.[1] And yet, flows across the region never stopped; they merely slowed down, changed in character, made themselves invisible, or found other routes. The authors discuss these changes in the past and the present, and look at specific places and situations in order to address broader questions of how interconnections and flows have been mediated, facilitated or blocked. The political fragmentation of this region has always had an influence on mobility, also during times of imperial control. At times, it has worked to facilitate mobility that an imperial government sought to block, or produced regional conflicts that enhanced military mobility but complicated the same for civilians. Even so, the studies in this volume show that armed conflicts certainly created hurdles, but they were never able to prevent the movement of goods, animals, humans or ideas.

The chapters show that it makes little sense to assume rigid edges to this part of the trans-Himalayan space. Territorial borders certainly do not provide such edges. Firstly, borders have been extraordinarily mobile, unstable and permeable over time. Secondly, there is no reason to assume that unstable, permeable borders were prevalent in the past but have disappeared with the emergence of modern nation-states. It is misleading to apply a 'Westphalian logic' that takes sovereign nation-states as points of departure. It is true that the states administering the region established border controls, state-to-state agreements, and strict interdictions of cross-border mobility. But focusing on these measures obscures the fact that there are other layers of control that continue to facilitate movement of goods and people across national borders. The chapters highlight how flows negotiate this multi-layered political landscape across time. The authors engage with the movement of people and everything that people bring, from goods to ideas. They focus on several commodities, especially jade, capital, drugs, guns and animals. This allows us to compare flows within and across a limited region over time, as well as flows between locations far apart at a single point in time.

1 This book focuses on the easternmost part of the Himalayas (which we call the India-China Corridor), not on the entire Himalayan massif. It seeks to contribute to discussions about flows and spaces, multiple state space, and multistate margins – and to link these to two emerging scholarly fields, the study of Asian borderlands and Trans-Himalayan studies. On multiple state space, see Shneiderman, 'Central Himalayas', pp. 289-312. On multistate margins, see Smyer Yü, 'Introduction', pp. 11-41. For a recent enrichment of the literature on the eastern Himalayas, see contributions to Smyer Yü and Dean, eds., *Yunnan-Burma-Bengal Corridor.*

The volume is a conversation between historians and anthropologists who work at the crosscurrents of Asian borderland studies and Trans-Himalayan studies. Their contributions allow us to reflect on evolving relationships of time, space and place, as well as on the importance of combining the ethnographically historical and the historically ethnographic. In line with many recent studies, the volume challenges the conventional foregrounding of nation-states (a short-lived phenomenon in the longer view[2]) without, however, resorting to the fantasy of disembodied flows. The volume also asserts that contemporary events require long-term perspectives in order for us to understand how past mobilities underlie many of today's complex conflicts. With an ethnographic eye on places, we can analyse how flows of ideas and goods can alter the meanings of those places, ideas and goods. People do not have to travel themselves to be influenced by movement. Anchored in the region's stunning landscapes (but not determined by them), flows are negotiated in webs of human interaction (but not defined by them).

A corridor between India and China

In the research collaboration that has supported the conversations of which this volume is an outcome, we have used the word 'corridor' as a marker to frame mobility and flows.[3] The term encompasses fluctuating and shifting – even volatile – movements across an often simultaneously changing socio-political geography. Mobility marks and makes such a geography. Localities come to life or fade away with the energy of movement. Market towns emerge and disappear with the movement of goods and all attendant interactions. Refugee camps mushroom out of an acute situation, connecting places that are sometimes far apart, only to become permanent, disappear completely, or turn into townships. The corridor's geography is under constant reconfiguration.

As a geographical expanse, the 'India-China Corridor' (as we shall refer to it from now on) is best thought of as a broad swathe of land that stretches from the Bay of Bengal to Yunnan in China, crossing the eastern Himalayas and its southern promontory, the Indo-Burma Range. Thus, it is part of the

2 Using the nation-state to frame research is widely known as methodological nationalism. Wimmer and Glick Schiller, 'Methodological Nationalism', pp. 301-334.

3 'The India-China Corridor', including the team members Gunnel Cederlöf (P.I.), Willem van Schendel, Mandy Sadan, Dan Smyer Yü and Arupjyoti Saikia, funded by the Swedish Research Council 2016-19.

eastern trans-Himalayan space of exchange, but it does not cover the entire Himalayan massif. Interaction has a long history here and social life has never been fixed. Tharaphi Than and Htoo May show this clearly in their chapter on Sittwe in Rakhine State, Myanmar. When citizenship came to be defined by ethnicity and religion, long-time citizens found that they had been re-categorised and were forced to move into 'immigrant' townships. Emphasising mobility in a study of a particular locality does not imply that we have reservations about people's sense of being rooted in a place. Rather, it highlights our point that, even in places associated with lengthy histories of belonging, we must take account of how flows of people, things and ideas continually shape and change places. Generations of people may live in a particular place, but they also travel, return, and receive visitors. Influences from afar are forever transforming any place.[4]

Countless studies with roots in nineteenth-century ethnography have assumed that the people of the 'India-China Corridor' are best understood as a smorgasbord of isolated and place-bound cultures. Following the external and racialised views of colonial explorers and conquerors, they have been represented as 'primitives' and 'tribals.' Colonial ethnographers tended to label people living outside the colonially administered territories and refusing to negotiate or yield to power as 'wild' and 'uncivilised.'[5] People placed into these categories were either praised for their assumed closeness to nature and simple customs, or judged for their so-called backwardness and lack of development. What followed was a century of ethnographic studies that focused on single 'tribes' and their supposedly narrow worlds. These studies contributed to a long tradition of obscuring their mobility and connections with the wider world. Such biased wisdom circulated widely (Figure 1.1) and continues to shape outsiders' popular views as well as state policies. In her chapter, Joy Pachuau details such encounters and processes for people in the Lushai Hills (now Mizoram) in Northeast India.[6]

Each chapter, in its own way, demonstrates the continuously shifting interaction of places and mobility. And in every encounter, there is friction. In Anna Tsing's words: 'Friction is not just about slowing things down. Friction is required to keep global power in motion.' For example, in the India-China

4 Than and May, chapter 9 in this volume.

5 For example, a map of the 1850s has this legend over the area now known as Chin State (Myanmar) and Mizoram (India): 'Unexplored Mass of Hills inhabited by a great many Tribes of Indo-Chinese type, in low stages of civilization.' In the accompanying text, the area is described as 'inhabited by a vast variety of wild tribes of Indo-Chinese kindred (known as Kookees, Nagas, Khyens, and by many more specific names.' Yule, 'Geography of Burma', pp. 54-108, at 54 and 65.

6 Pachuau, chapter 5 in this volume.

Figure 1.1 Photograph captioned 'Group of Wild Chins'[7]

Corridor, friction works as a double-edged sword in forging new overland connections. On the one hand, the construction of new roads (often for extractive or military purposes) created new pathways that made motion easier and more efficient. On the other hand, they were attempts to channel

7 Photographer unidentified and date unknown (early twentieth century). Credit: Ethnologisches Museum, Staatlichen Museen zu Berlin.

movement and control where people and goods travelled.[8] The chapters in this volume show that such 'efficient' roads may trigger the use of other pathways, away from public sight and across harsh terrain, to make sure that items – from daily commodities to guns and drugs – reach otherwise sealed-off places. They show how friction produces new forms of mobility.

Geography as destiny?

The India-China Corridor is a truly ancient passageway. Current evidence points to the arrival of the earliest humans here at least 40,000 years ago. They came from the west, moving along the southern margins of the Himalayas and along the northern coast of the Bay of Bengal. Some settled in the richly biodiverse environments of what are now Bangladesh and Northeast India; others pushed on – across the eastern Himalayas and the Indo-Burma mountain range, via the marshy lowlands and into the hills and high plateaus, ultimately peopling Southeast and East Asia.[9] Ever since, mobility has been the hallmark of these trans-Himalayan spaces, and yet, scholars have often downplayed it in favour of territorialised perspectives, framed by the geographical contours of the – historically very recent – nation-states.

A new generation of researchers is now actively involved in overcoming this handicap. This is easier said than done because most of us were trained to foreground national frameworks and fixed entities and identities. Academic cultures, reference circles, and career paths also remain remarkably hidebound and compartmentalised. Had the authors of this volume stayed within such frameworks they would never have met, being based in the separate Area Studies domains of South, Southeast, and East Asia studies. Geography can be academic destiny. To help the process of deframing this space (in both spatial and disciplinary terms) and to suggest how to reframe it in multiple ways, the 'India-China Corridor' research programme has organised meetings in this region to bring together scholars working on flows and frictions. The chapters in this volume are one of the outcomes of these meetings. They show how the early-modern to modern periods, especially, have been shaped by human interactions in and between places that we now know as Assam, Nagaland, Kachin State, Rakhine State, and the Shan-Dai borderlands of Myanmar and Yunnan. All demonstrate our need to understand flows and frictions as perpetually joined, as partners

8 Tsing, *Friction*, especially pp. 5-6.

9 Bae et al., 'On the Origin', pp. 1-7.

in an endless dance. But who leads? David Ludden emphatically makes the point that mobility always precedes attempts at regimenting space by means of territorial control. Social space is essentially mobile space, and territorialisation of social power is often a brutal intervention to regulate it.[10] With different takes on the question, the authors show how mobility and territorial control have been dynamically and differently intertwined. Flows may alter or circumvent spatial control, and regimented space can cause multiple identities to grow and move places.

Flows are never smooth, and mobility always comes with obstacles. From the Indo-Burma mountain range to the Yunnan plateau, and from the Indian Ocean to the Himalayas, a traveller will need to cross huge monsoon-swollen rivers at sea level and face thousand-metre climbs. The friction of terrain is very palpable – the terrain itself has always been a challenge to anyone travelling here. This includes areas where humans hardly ever set foot: the Himalayan glaciers. These are instrumental in providing the conditioning ecologies and seasonal variations that shape mobility. The climate, the seasons, and the ecozones have put pressure on human livelihoods and activity. As a result, researchers still routinely assume that geography is destiny, and that verticality has always constrained mobility. This assumption is refreshingly challenged in the chapters of this book, both in those that reach back centuries and in those that focus on recent times. The historical perspective on flows and frictions is developed in chapters by Jianxiong Ma and Gunnel Cederlöf (connecting the Shan-Dai territories with Burma and Qing China), Joy Pachuau (on the long-term mobility that influenced the Lushai Hills), and Arupjyoti Saikia (on Assam's connections with neighbouring trade networks). Contemporary flows and frictions are highlighted in chapters by Jelle Wouters (on Nagaland-Bangladesh connections), Tharaphi Than and Htoo May (on migrants in Rakhine state (Myanmar) and Henrik Kloppenborg Møller (on jade connections between the Kachin and Shan states in Myanmar and international markets).

It is clear that during the past century and a half 'distance-demolishing technologies'[11] have brought about considerable time-space compression.[12] That said, we should resist three common assumptions. The first is that previously mobility was obstructed. There may have been occasional blockages

10 Ludden, chapter 2 in this volume.

11 '[D]istance-demolishing technologies – railroads, all-weather roads, telephone, telegraph, airpower, helicopters, and now information technology ... changed the strategic balance of power between self-governing peoples and nation-states [and] diminished the friction of terrain.' Scott, *Art*, p. xii.

12 Harvey, *Condition*.

but, mostly, travelling simply took longer. Just think of the centuries-old transportation system based on mule caravans moving between Yunnan, Sichuan, and Tibet.[13]

The second erroneous assumption is that time-space compression happened across the board. On the contrary, it was more noticeable in some parts of the India-China Corridor than in others. Moreover, that which allowed a speedy passage for some people may have caused obstacles to others. Whether in colonial times or today, the construction of roads and railways serves the purposes of those who run the state. Moving along a highway or railway will definitely shorten your travel time, but only if it takes you to the places where you want to go and if you can pass the checkpoints and pay the fees. With such frictions, many prefer the long, winding and poorly maintained routes that also allow access – but only for those who can locate the paths or enter the social networks that will open them up. In other words, time-space compression is related to social power. In his chapter, Møller speaks of opaque networks of illicit trade that include marriage relations. He discusses the friction of bribes, taxes, road quality, and war. Ma and Cederlöf found that much the same applied in the same territories 150 years earlier. At that time, flows across the difficult terrain were conditioned by local kinship relations and political control through ties to the Burmese Court and the Ming and Qing bureaucracy in China. Flows were also considerably structured by travellers' linkages to cross-border merchant networks. Knowledge about the places and routes where time-space compression was more favourable was, and still is, a well-guarded domain.

The third mistaken assumption is that time-space compression has been a linear process that accelerated over time, spurred on by capital and technology. In reality, its pace fluctuated. For example, it advanced suddenly and rapidly during the Second World War, with new airfields and metalled roads, only to retreat in the aftermath, leaving a rash of abandoned, overgrown airfields and neglected, potholed roads.[14]

The extremely varied geography, ecology and climate that characterise the India-China Corridor make their mark on social life and spatial interaction. Uplands, high mountain ranges, and deep ravines structure the pathways that connect villages and towns, irrespective of their location. In Nagaland, towns are on mountaintops, and in the Shan-Dai region they are in the flat basins and surrounding hills. Travelling in the lowlands includes negotiating

13 Ma and Ma, 'Mule Caravans', pp. 24-42.

14 Cederlöf, chapter 4, Møller, chapter 8 and Ma, chapter 3 in this volume.

rivers and marshes that seasonally swell from snow melting in the Himalayas and from monsoon rains. The rugged landscapes and the resulting friction of terrain of the India-China Corridor shape overland mobility in specific ways. Mountain passes, river courses, and valleys continue to guide flows. Despite today's modern infrastructure, long established network nodes and pinch points still play a part in structuring mobility. This occurs especially where old technologies of mobility persist, such as mule caravans, small boats, bamboo bridges across mountain streams or people carrying loads on their backs. Even though the reach of railways and motorable roads has increased, old transport technologies have not been marginalised as much as one might think. In many parts of the Corridor, roads (let alone railways) are few and far between, and people often avoid them out of fear of military presence, police or forest department checkpoints, or guerrilla attacks. The old routes and trails are still in use and effectively link people across the Corridor and well beyond. These old technologies embody a huge store of knowledge about how to navigate mobility and territorial control.[15] Magnus Marsden argues that these pathways must not be treated as:

> ... informal and illegal and thereby inevitably a security threat or a risk. They are better thought of as monuments to the creative activity of people who have been poorly served by the nation-state and the international system over the past decades. It is in this context that they have built their own infrastructures, both for life and for commerce.[16]

Thus, pathways and the infrastructures that sustain them are more flexible than the landscapes that they traverse. In the India-China Corridor, geography is highly relevant, but not destiny.

The production of ignorance

Scholars and policy makers have rarely understood the dynamics of the India-China Corridor. Generally speaking, they have been dazzled by its

15 According to Martin Saxer, a pathway can be understood as 'a configuration that is at once geographical and social. A pathway is thus not just another word for trade route [...] Life along a pathway is shaped by things, stories, rumors, and people passing through – by motion, or by flows, if you will. However, a pathway is neither just another word for flow. While shaped by motion, pathways are also conditioned by terrain, infrastructure and environmental factors like climate and weather.' Saxer, 'Pathways', p. 105.

16 Marsden, 'Actually', p. 30.

remarkable cultural diversity and understood it as highly fragmented. It is true that the India-China Corridor is a linguistic, cultural and religious shatter zone with only a few counterparts in the world, such as the Caucasus or New Guinea. The Corridor is home to hundreds of ethnic communities speaking dozens of languages belonging to a handful of language families. Many consider themselves to be Buddhists, Christians, Muslims, Jews, or Hindus, and many others observe local community religions. They do not share a single link language. In some of the sub-regions of the Corridor national languages (Chinese, Burmese, Bengali and Hindi) are relevant as link languages, whereas in other sub-regions communication is facilitated by specific Corridor languages such as Shan, Jinghpaw, Lushai and Arakanese, as well as creole or pidgin languages, notably Nagamese.

However, it would be a mistake to overplay the extreme ethnic and linguistic fragmentation when studying mobility and territorial power. When such fragmentation is emphasised, it is usually seen to be a result of the Corridor's craggy landscape and a historical absence of large, enduring, hegemonic states. Nevertheless, this appears not to have hampered mobility. As Cederlöf shows, there were multilingual traders and interpreters who provided services along the pathways. Moreover, webs of mobility were not ethnically exclusive and instead, as Ma and Møller show, they were and remain mostly highly multi-ethnic. The specific forms that these mobility-webs take today are historically specific.

What cultural fragmentation did achieve, however, was that it made the India-China Corridor very hard for outsiders to access and understand. They could observe people and goods entering into and emerging from it, but its inner workings remained largely inscrutable. When we follow British travellers' trails in the early nineteenth century, as Cederlöf does, reading their reports is like seeing pieces of scenery beyond which the travellers were unable to reach. Of course, their observations could be detailed and accurate, but they were fragmentary and frequently led to inaccurate conclusions. These British visitors listened to rumours and picked up impressions from conversations with people on the road. They followed these up with long internal discussions about what the information, silences, and absences might mean. The observations written into their reports and correspondence included events that the officers failed to understand. These open up a deeper understanding of interactions and social life along the pathways and in the basins, when read against internal administration, local genealogies and natural science data.

It should not surprise anyone that many outside observers imagined the Corridor and its inhabitants to be mysterious and outlandish. Stories about tree-living creatures, complete nudity, human sacrifices, headhunting and fabulous natural riches circulated widely in China, India, and Burma alike. Echoes of these ideas still shape contemporary notions of the region, and there is a need to decolonise such perceptions. Significantly, they point to an important aspect in the study of mobility and territoriality: the production of ignorance. As Møller explains, ignorance is not just the absence of knowledge, or an effect of the limits of knowledge. It can also be deliberately produced by guarding knowledge against outsiders. He shows how Chinese jade traders do this by hiding the harsh realities of jade extraction in Myanmar behind a smokescreen of positive imagery about the beauty of jade and its cultural significance in Chinese history.[17]

People on the move, whether as migrants or people with itinerant livelihoods, often attract negative attention. Here, too, the control of information flows is highly relevant to mobility. States prefer a settled and taxable population, even targeting particular elusive groups as lawless, as in the 'criminal tribes' in British-administered territories in Asia. Rumours about migrants raping and stealing the belongings of honourable citizens can spread like wildfire. Labour migrants from Bangladesh arriving in Nagaland have found themselves in a squeeze between what Wouters describes as an escape from endemic poverty, devastating floods, and acute land shortages in Bangladesh. When settling in Nagaland, they find that the Naga political struggle for an ethnic homeland turns notions of Christian identity and ethnic purity against migrants. In 2015, the president of the Naga Students' Federation claimed that 'immigration from Bangladesh has become a serious threat to Naga society,' and a month later a furious mob attacked and lynched an evidently innocent Assamese man (whom they mistook for a Bangladeshi Muslim), accusing him of having raped a Naga girl.[18]

As new technologies become available, new research methods are developed to challenge the active production of ignorance. This is demonstrated in Than and May's chapter, which relies heavily on digital proximity to overcome physical barriers. It was impossible for Than to meet up with her Rohingya interlocutors whose mobility was severely constrained by state action. The Myanmar authorities isolated them from the general population by issuing separate identity cards and driving them into camps. This was a strategy that nurtured ignorance about their condition, made them invisible and voiceless, and strengthened what Than and May call 'mental borders.' By using digital connectivity, their Rohingya interlocutors were able to

17 Møller, chapter 8 in this volume.
18 Wouters, chapter 7 in this volume.

overcome their physical immobility and imposed invisibility. This enabled their ideas and concerns to escape the camp. In this way, hidden histories and unacknowledged experiences became visible.[19]

The fact that the India-China Corridor has been hard for outsiders to understand has resulted in scholars marginalising it in their work. More concerned with explaining how power, ideas and wealth radiate from 'centres' to 'peripheries,' they have often failed to acknowledge the powerful impact that this hub of connections has had on 'centres.' An exception is research into the large-scale impact of flows of opium and methamphetamine, produced in the eastern parts of the Corridor (dubbed the Golden Triangle) and supplied around the world. The impact of other Corridor flows – of industrial commodities, gemstones, fossil fuels, people and ideas – on 'centres' outside the Corridor is much less studied, and the same is true for power centres within the Corridor.

The India-China Corridor's marginalisation in scholarly imaginations has hampered absolutely essential debates and cooperation. Only very recently has a scholarly field begun to take shape in this world region. The obstacles to be overcome are considerable. Firstly, as in Than and May's work, we have seen that physical access is not always possible. For security reasons, states severely restrict free movement in parts of the Corridor where there are local revolts and border issues. Secondly, researchers may have only limited or selective access to historical source material in state archives. The archives may be closed, or the necessary permits may be hard to obtain. Thirdly, it is an obvious hurdle in such a multilingual region that individual researchers have limited language skills. This means that they cannot access information in many different languages unless through interpreters or translators. As a result, they are obliged to concentrate on sources that fit their linguistic profile. Thus, Ma gravitates towards Chinese, Cederlöf to English, and Than and May to Burmese. Fourthly, this is a new academic field. Novel conceptual tools for studying the Corridor, such as Southeast Asian Massif, Zomia, borderworld, multiple state space, and multistate margin, have only recently been developed and tested.[20] Each comes with its own discussions that overlap, but do not coincide, with the focus of this volume. Studying the Corridor requires a specific form of teamwork and the pooling of specific language skills. Developing expertise in handling the sources of information is especially important if we are to recognise biases and silences in the sources themselves as well as in their interpretation. Such expertise involves an

19 Than and May, chapter 9 in this volume.

20 Shneiderman, 'Central Himalayas', pp. 289-312; Sadan, *Being and Becoming*; Smyer Yü, 'Trans-Himalayas', pp. 8-41.

understanding that documents which are kept in colonial archives do not simply reflect imperial mind-sets. On the contrary, many archival records are, in fact, letters and reports produced by Burmese, Shan, Assamese, and other non-Europeans. Imperial rulers depended on access to vernacular documents and intelligence, which explains why many such documents are now located in what tend to be called 'colonial archives.' It would be a huge mistake to discard these vernacular or subaltern voices as representing 'colonial bias.'

Research cooperation is essential to reconnect the study of the India-China Corridor. It can break down barriers, repair fragmentation and overcome the state-centred perspectives that have long impeded and distorted research. It can also give an impetus to decolonising, departitioning and demarginalising studies of this large and neglected region. It is our hope that this volume can contribute to this endeavour.

Organisation of the book

This volume consists of three Parts. *Part One: Prologue* contains David Ludden's wide-ranging essay on ways to reconceptualise the spatial history of Asia. It can be read as a thought-provoking framing of the issues involved in rethinking mobility and friction.[21] *Part Two: A Long View* consists of four chapters by Jianxiong Ma, Gunnel Cederlöf, Joy Pachuau and Arupjyoti Saikia. Each focuses on historical flows and frictions that manifested themselves in a particular location during a well-defined period. Together, these chapters suggest new ways of studying Corridor mobilities across several centuries. The volume concludes with *Part Three: Mobilities Today.* Made up of three chapters by Jelle Wouters, Henrik Kloppenborg Møller and Tharaphi Than and Htoo May, it considers the interplay of flow and friction in three forms of contemporary mobility in the Corridor – of people, commodities and identities – and their powerful effects on individuals and communities.

References

Bae, Christopher J., Katerina Douka, and Michael D. Petraglia. 'On the Origin of Modern Humans: Asian Perspectives'. *Science* 358, eaai9067 (2017).

Harvey, David. *The Condition of Postmodernity: An Enquiry into the Origins of Cultural Change*. Cambridge, MA and Oxford: Blackwell, 1990.

21 See also Ludden, 'India's Spatial History', pp. 23-37.

Ludden, David. 'India's Spatial History in the Brahmaputra-Meghna River Basin'. In *Landscape, Culture, and Belonging: Writing the History of Northeast India*, edited by Neeladri Bhattacharya and Joy L.K. Pachuau, pp. 23–37. Cambridge: Cambridge University Press, 2019.

Ma, Jianxiong, and Cunzhao Ma. 'The Mule Caravans of Western Yunnan: An Oral History of the Muleteers of Zhaozhou'. *Transfers* 4, no. 3 (2014): pp. 24–42.

Marsden, Magnus. 'Actually Existing Silk Roads'. *Journal of Eurasian Studies* 8, no. 1 (2017).

Sadan, Mandy. *Being and Becoming Kachin: Histories Beyond the State in the Borderworlds of Burma*. A British Academy Postdoctoral Fellowship Monograph. Oxford: Oxford University Press, 2013.

Saxer, Martin. 'Pathways: A Concept, Field Site and Methodological Approach to Study Remoteness and Connectivity'. *Himalaya – The Journal of the Association for Nepal and Himalayan Studies* 36, no. 2 (2016): pp. 104–119.

Scott, James C. *The Art of Not Being Governed: An Anarchist History of Upland Southeast Asia*. New Haven, CN and London: Yale University Press, 2009.

Shneiderman, Sara. 'Are the Central Himalayas in *Zomia*? Some Scholarly and Political Considerations Across Time and Space'. *Journal of Global History* 5, no. 2 (2010): pp. 289–312.

Smyer Yü, Dan. 'Introduction: Trans-Himalayas as Multistate Margins'. In *Trans-Himalayan Boundaries: Livelihoods, Territorialities, Modernities*, edited by J. Michaud and D. Smyer Yü, pp. 11–42. Amsterdam, Amsterdam University Press, 2017.

Smyer Yü, Dan. 'Trans-Himalayas as Multistate Margins'. In *Trans-Himalayan Borderlands: Livelihoods, Territorialities, Modernities*, edited by Dan Smyer Yü and Jean Michaud, pp. 8–41. Amsterdam: Amsterdam University Press, 2017.

Smyer Yü, Dan, and Karin Dean, eds. *Yunnan–Burma–Bengal Corridor Geographies: Protean Edging of Habitats and Empires*. New Delhi: Routledge, 2021.

Tsing, Anna Lowenhaupt. *Friction: An Ethnography of Global Connection*. Princeton, NJ: Princeton University Press, 2005.

Wimmer, Andreas, and Nina Glick Schiller. 'Methodological Nationalism and Beyond: Nation-State Building, Migration and the Social Sciences'. *Global Networks* 2, no. (2002): pp. 301–334.

Yule, Henry. 'On the Geography of Burma and its Tributary States, in Illustration of a New Map of those Regions'. *The Journal of the Royal Geographical Society of London* 27 (1857): pp. 54–108.

About the authors

WILLEM VAN SCHENDEL, University of Amsterdam and International Institute of Social History, the Netherlands. He works with the history, anthropology and sociology of Asia. Recent works include *A History of Bangladesh* (2020), *Embedding Agricultural Commodities* (2017, ed.), *The Camera as Witness* (2015, with J. L. K. Pachuau). See uva.academia.edu/WillemVanSchendel.

GUNNEL CEDERLÖF, Professor of History at Linnaeus University, Sweden, and member of the Centre for Concurrences in Colonial and Postcolonial Studies. She studies environmental, legal, and colonial history in India and South Asia. Publications include *Founding an Empire on India's North-Eastern Frontiers, 1790-1840* (2014), *Landscapes and the Law* (2008) and *Ecological Nationalisms* (2006 with K. Sivaramakrishnan).

Prologue

2 Spatial History in Southern Asia

Mobility, Territoriality, and Religion

David Ludden

Abstract

Flows and frictions in mobile spaces connecting South and East Asia occupy spatial histories in monsoon environments that span the arid northern steppe and the humid southern tropical seas. Many centuries of migration, trade, travel, and settlement along routes across the Hindu Kush, up and down the Indo-Ganga Basins, and around Indian Ocean coastal regions produced expanding cultural spaces, where mobile warriors and local elites formed strategic alliances and power relations that turned social space into political territory. Patronage for religious rituals, elites, and knowledge served to sanctify these power relations. Religion thus became a territorial technology, and its powers of social boundary-making became more strident during the mid-twentieth-century shift from more fluidly inclusive imperial territory to more rigidly exclusionary nation-state territorialism.

Keywords: Indo-Persia, Mongols, Indian Ocean, monsoons, religions

SARS-Cov-19[1] and Black Lives Matter (BLM)[2] have fired up my thinking about spatial history.[3] The travels of the virus dramatize the power of uncontrolled mobility in shaping and disrupting human environments where territorial authorities seek to maintain orderly control over mobility with programmes of spatial containment. That interaction of mobility and territoriality has appeared in news of refugee crises, immigration politics, and globalisation

1 It has that name because of its eighty per cent genetic similarity to the Severe Acute Respiratory Syndrome (SARS) coronavirus and its identification in December 2019, Covid-19

2 https://blacklivesmatter.com/

3 This chapter builds on my keynote presentation for the November 2019 Stockholm conference, 'Flows and Frictions in Trans-Himalayan Spaces'.

Cederlöf, Gunnel, and Willem van Schendel (eds), *Flows and Frictions in Trans-Himalayan Spaces: Histories of Networking and Border Crossing*. Taylor & Francis Group 2022
DOI: 10.5117/9789463724371_CH02

debates, but the virus has made it a matter of life and death. At the same time, and waged amidst disproportionate mortality among poor people of colour, BLM struggles against the violence that infects state territorial control and has worsened – as we will see – during the speeding globalization that spawned the pandemic; though framed in an American territorial context, BLM has spilled out uncontrollably into global spaces of racial inequity and social activism.

Spatial history

Covid-19 and BLM thus highlight the utility of spatial history for interpreting the present. My task is to explore spatial history in Asia.[4] In that light, present-day violence against African-Americans can be seen as part of the historical space of trans-Atlantic plantation slavery operating inside the globalization of racial capitalism, all around the world, embracing Asia.[5]

The pandemic inhabits much older space-time, still more deeply rooted in Asia. Plagues sailed along with spices on monsoon winds across the Indian Ocean to Europe for twelve centuries before Mongols built overland routes that the Black Death travelled from Mongolia to Crimea, sailing from there all over Europe; that whole trip took thirteen years.[6] Five centuries later, cholera went much further at the same pace, when the British Empire brought the world our first truly global pandemic, in 1817: the cholera bacillus took fourteen years to travel from Bengal to England, and two more to sail to America.[7] Industrialization sped up pandemics (and climate change) dramatically, so that in 1918, the flu virus came on road, rails, and steamships from Kansas to Europe to India in two months, and in 2019, SARS-Cov-19 jetted from China to Iran, Italy, and the US in a few weeks.

Spatial history thus begins with mobility. As Michel de Certeau says, 'space is composed of intersections of mobile elements.' Social space is composed of moving elements (including people, ideas, animals, art, water, commodities, skills, poetry, and viruses) which intersect on sites of social power. Each site becomes what de Certeau calls a 'place,' forming an orderly arrangement of

4 Ludden, 'Maps', pp. 1057-1078.

5 Manjapra, *Colonialism*.

6 Note that Marco Polo only needed three years to travel by land from Italy to China, 1271-1274, indicating that infectious disease travels more slowly than the fastest long-distance travellers, because it has to move randomly from one traveller to another.

7 Aberth, *Plagues*.

things in space, which 'implies an indication of stability.'[8] A place, a route, or network among places, and any coherent collection of places form what I call 'territory.' Social power relations form places and provide indications of stability for territory – notably with boundary maps – but can never completely overcome the disorderly mobility that forms social space. In short, territory is a project of social power striving to organize, direct, or contain mobility, and in this essay I consider the idea that religion can be a territorial technology.

I think of the 'India-China corridor' of 'flows and friction' as a territorial metaphor that organizes a mobile space conceptually. Routes through it carried cowry shells from Bengal to Yunnan and troops along the World War Two Ledo Road; now it contains mobile space for the Belt-Road Initiative, and connectivity among its disparate peoples-in-places forms a major theme in this volume. However, as several authors also note, unruly mobility dissolves the corridor's territorial coherence: a long list of historically relevant mobile spaces flow out of it, following rivers, migrations, languages, communists,[9] ethnic conflicts,[10] supply chains (for guns, drugs, and other commodities), and now, the coronavirus.

In addition, in the long term, India and China are unstable territorial signifiers. Repeated armed conflicts today on western and eastern edges of southern Tibet indicate that these two imperial states have drawn territorial maps in mobile space that neither controls. These conflicts erupt today in spatial histories that spill across West, Central, East, and Southern Asia. The India-China corridor thus inhabits a vast spatial history; it forms a tangled intersection of routes, where Brahmaputra, Barak, Chindwin, and Irrawaddy rivers glide through mobile imperial space that only became a 'space between India and China' when Qing and British empires mapped territorial borders around Burma.[11] I approach the corridor via spatial histories of southern Asia.

Indo-Persia

Mobility in southern Asia travels monsoon environments. Every year, winter winds blow frigid and dry out of the arid, northern Eurasian steppe, surging into warmer southern humid tropics, cooling the air, drying the land, and pushing sailors at sea from north to south and east to west. Summer winds

8 See de Certeau, *Practice*, pp. 115-130; quotes are from p. 117. He also introduced the idea of spatial storytelling.
9 Sacks, 'Strategy', pp. 227-247.
10 Reliefweb, 'Conflicts'.
11 Schwartzberg, *Historical Atlas*, p. 56.

blow in the opposite direction, raging inland soaked with rain from tropical seas, heating the air, watering lands below the Himalayas, pushing sailors at sea south to north and west to east.

The Himalayas are drenched with monsoon rains that pour into great rivers flowing out of Tibet into lowlands of South, East, and Southeast Asia. In their northern rain shadow, arid deserts and semi-arid grasslands have nurtured nomad cultures from prehistoric times, where highlands send rain and snowmelt into oases that became strategic sites for travellers on the network of animal tracks, ruled by nomads, which came to be called Silk Roads. Travellers across that vast dry land of cold winters and hot summers – stretching from Manchuria to Ukraine and Persia to China – produced mobile spaces that shaped all Asian societies. Many of them travelled south into the tropics, seeking greener pastures, as the southern coast also received many settlers who travelled the Indian Ocean, who set off to the east in summer, returned to the west in winter, and often settled along the coast in the east, in Gujarat, Malabar, the Maldives, Sri Lanka, Arakan, the Malay Peninsula, Southeast Asia, and south China.

South Asia thus became the crossroads of Asia, where travels by land and sea came together on a vast land bridge connecting Central Asia with the Indian Ocean, studded with seaport coastal homelands for overland and overseas settlers. Settlers came in constant trickles and periodic major waves from the north, west, and east. All three mobile spaces met, in the eighteenth century, in the Meghna River basin, where Afghan, English, Greek, and Khasi warrior merchants haggled in mountains above Sylhet on routes to Yunnan through 'the India-China corridor.'

We can call that land bridge Indo-Persia. Nature shapes its mobility. Complex patterns of natural resource distribution foster trade among its ecologically diverse environments. Herds and herders move to greener pastures: they often head to the hills, out of parched plains, in summer and return to lowlands in winter. Hunters, gatherers, nomads, fisher folk, and shifting cultivators have moved around constantly over the centuries, as struggles for land, labour, and wealth propelled people into open frontiers. Mountains also filtered mobility from the steppe: nomads could ride directly into fertile Persia and China, with ease of entry and exit, but they had to cross the Hindu Kush to reach southern Indo-Persia. Those mountains became a maze of capillary filters for people travelling in all directions; many settled down to work the mountain landscape, raise animals, tend crops, and pursue mercantile and military careers in mobile spaces connecting Persia and the steppe with Tibet, Kashmir, Afghanistan, and Punjab. Migrants travelling south from the steppe rarely returned: they intermarried, started families, and merged into older local populations throughout Indo-Persia.

Spatial history in southern Asia thus includes a vast scattering and mingling of peoples on landscapes that remained wide open for travelling settlers until the late nineteenth century. Those images that we inherit of South Asia being an ancient land of static enclosed village communities were modern imperial instruments of territorial administration. People actually lived in places of many shapes and sizes, changing over time, forming urban clusters along trade routes and rivers, and, by the seventeenth century, hosting manufacturing centres that produced cloth for domestic markets and world trade, as well as many products for local consumption. Urbanism most typically took the form of complex clusters of small dense settlements, strung along trade routes. Cities began attracting more migrants with expanding military fortification, commerce, and manufacturing after 1500.

Available evidence indicates that, in the eighteenth century, villagers rarely engaged only or even mostly in sedentary farming; monsoons made sure of that, pushing people off farms with floods and droughts, moving workers to plough, plant, and harvest here and there, and leading farmers into open frontiers and many occupations. Farming was intricately interwoven into networks of mobility, trade, alliance, and conflict, in loosely organized agrarian regions composed of urban centres, towns and villages large and small as well as countless tiny satellite settlements, many mobile, seasonal, and temporary, to accommodate workers moving to live where work was available.

Modern boundaries dividing Asian regions are also instruments of modern territorialism. They construct spatial limits and structures for mobility that were carved into spatial history by nineteenth-century empires. It is more realistic to envision historically dynamic social spaces where people travelled and settled everywhere on a vast land bridge connecting Central Asia and the Indian Ocean, where travellers and traders routinely followed routes across West Asia, as did cholera after 1817, walking from Bengal through Afghanistan to the western steppe and sailing from there by sea, like the Black Death, to England.

The expansive spatial history of Indo-Persia eventually embraces all of what are today Iran, Uzbekistan, Turkmenistan, Tajikistan, Kyrgyzstan, Afghanistan, Pakistan, India, Nepal, southern Tibet, Bangladesh, and western Myanmar, spilling across the Indo-China corridor. Its recorded history begins in the second millennium BCE, with the *Vedas* and the *Avesta*, two religious texts composed in two branches of the Indo-Aryan language family that was shared by related populations of nomads whose mobility formed social space spanning the western steppe, Persia, and the Indo-Gangetic basin. Durable evidence of religious thought, elites, and rituals indicate the mobility of Indo-Persia as social space from the outset.

Ancient non-literate Indo-Aryan language-speaking horse-herding warrior nomads travelled territories they formed in northern and southern Indo-Persia. Organized imperial territories emerged when Achaemenid Persia (700 to 330 BCE) occupied the Indus Valley on routes down the Ganga Basin into territories called *Janapatha* and *Mahajanapatha*. Summertime routes through the Hindu Kush carried traffic in all directions, but sadly for Alexander the Great, steppe nomads forced his armies south across the Hindu Kush in winter. His desperate troops revolted, and he fled down the Indus back to Babylon after romping victoriously through Persia. His journey marks an early stage in the interlacing of territories across Indo-Persia, from Silk Roads to Kanya Kumari, which became increasingly intertwined in centuries that followed.

Territorial transitions

Buddhists provide good documentation and architectural evidence of histories of spatial transformation. Beginning soon after Alexander's journey, monks carried the Buddha's *Dhamma* in all directions from the eastern Ganga Basin: southwest along river basins, deep into the southern peninsula into Sri Lanka; southeast into Arakan and Burma; northwest through the Hindu Kush into Persia, and from there east across the Silk Roads into China, Japan, and Korea. After 250BCE, monks, monasteries, stupas, temples, and rituals received imperial patronage from the Mauryan emperor Ashoka, who launched a burst of mobility that sent Mahayana Buddhism to China in the space of a century, by which time there was already a thriving Theravada Buddhist kingdom at Anuradhapura in northern Sri Lanka.

Patrons of Buddhism made religion a territorial technology. Their scale of operations expanded dramatically with increasing imperial territory during the Classical Age, which spans days of the Mauryas (322-185BCE) and Guptas (319-467CE) in the south, Parthians (247 BCE-224CE) and Sassanids (224-651CE) in the northwest, and the influential connective mobility of Indo-Aryan nomadic Kushanas and Indo-Greeks in the north, for roughly six hundred years after Alexander's departure.

Buddhist architecture proclaimed growing commercial wealth at strategic places strung like pearls along routes connecting China, Persia, Bengal, South India, Sri Lanka, and Arakan, where warriors and merchants formed patronage partnerships using wealth from trade to make donations and endowments for monks, monasteries, and temple sites of ritual and pilgrimage.[12] Rulers supported merchant communities in order to attract settlers and to generate local urban economies in places of refuge, protection, and capital accumulation.

Monuments proclaimed places with territorial identities as imperial icons and wonders of the world, starting with Ashoka's grand stupas, edicts, and pillars. In the sixth century, Kushanas sponsored the massive Buddhas south of the Hindu Kush in the highland oasis town of Bamiyan, gigantic statements of their devotion and major landmarks guiding travellers along Silk Road routes connecting Sassanid and Gupta territory.[13] Lavish monasteries arose in the nearby market town of Ghazni, south of Kabul on the main north-south route through Kandahar.[14]

A major transformation in the composition of imperial territory began in the sixth century, as patterns of patronage shifted during waves of warrior nomad migration, driven by Turkic language speakers from Mongolia, who crashed into agrarian empires from Europe to China and Indo-Persia. In that context, religious patronage became more militant. In Persia, Sassanids made Zoroastrianism the state religion. Guptas institutionalized Hindu Brahmanism. The fall of Rome set the stage for militant Christianity in Byzantium and Islam in Umayyad Syria.[15] In 751, Arab armies strengthened by Turks from the steppe drove the Chinese from the Talas River in the Central Asian Silk Road heartland; that further energized Muslim merchant, warrior, and missionary travels over trade routes and centres filled with Buddhists.[16]

Driven by steppe migrations, warriors and warrior-farmers poured into Indo-Persia: they were Turks from the eastern steppe, Arabs from the west, and nomadic Kushanas, Sakas, and Hunas from the north. Trade expanded in all kinds of commodities, including war horses. As a result, more militant warrior regimes emerged in post-Gupta-Sassanid centuries. Harshavardhana (Harsha) began this trend in the Ganga Basin, during fifty years of constant war (590-647) that made the region around Delhi a new fulcrum for imperial power for controlling war horse trade from the north and west and war elephant supplies from the east.[17] Harsha launched centuries of war that

12 Neelis, *Early Buddhist*.

13 Morgan, *Buddhas*.

14 Rienjang and Stewart, *Geography*.

15 Burbank and Cooper, *Empires*, pp. 61-92.

16 Elverskog, *Buddhism*.

17 Digby, *War-Horse*.

spread warrior power in all directions. Harsha reportedly began his reign with 20,000 cavalry, 5,000 elephant soldiers, and 50,000 infantry, but finished commanding 60,000 elephants and 100,000 cavalry.[18]

Harsha's patronage for religion was eclectic, poised as he was at the intersection of Hindu agrarian regions and trade routes filled with Buddhist merchant patrons. Most of his relatives were Hindu. He massively patronized Hindu temples, but he and his brother became Buddhists. He made connections with Silk Road Buddhists, routinely hosted monks at court, and continued Gupta patronage for the Buddhist centre of learning at Nalanda.[19] In the five subsequent centuries, however, royal patronage shifted to exclude Buddhists and favour Hindu deities to mark the rising authority of Brahmans in imperial territories of caste elite social power.

More militarised empires needed to mobilize more soldiers and more wealth from agrarian territories. A shift toward royal patronage privileging the local territorial authority of landed elites thus occurred as violent competition between Brahmans and Buddhists also increased, eventually sidelining Buddhists below the Himalayas. In the seventh century, a Brahman minister usurped the throne to oust a Jat Buddhist Rai dynasty in Sindh, and a Chinese Buddhist pilgrim bemoaned declining support for monks at Nalanda. A century later, a Tamil Brahman poet composed a lyrical account of Jains being massacred by Shaivas in Madurai. In the late eleventh century, a drama composed at the Chandela court vividly describes battlefield victories of Brahmans over Buddhist and Jain loyalists, a vision of Hindu warrior victories that became common in literature and folklore.[20]

By the eleventh century, Al Biruni – the first geographer to describe India as an imperial territory, who travelled south from Samarkand with Mahmud of Ghazni – concluded from local accounts that Brahmanical Hinduism defined India. Though Buddhism and Jainism still thrived around market centres, with private patronage, Hindu societies organized around Brahmans, temples, and Hindu *dharma* in *varna* ranks dominated major river basins where the major empires acquired control of land, labour, and capital. The last major dynastic patrons for Buddhism below the Himalayas were Palas in Bengal (750-1174), who had sent Tantric Buddhism into Tibet; they were followed by warrior Senas, staunch patrons of Brahmanism.[21]

18 Roy, *Military Manpower.*

19 Sen, 'Travel Records', pp. 24-33.

20 Verardi, *Hardships*, pp. 298ff.

21 Ludden, 'Decline', pp. 132-162.

Medieval to modern

The 'Sanskrit cosmopolis'[22] then formed a moving frontier of patronage for Brahman territorial privilege and authority with imperial alliances among warriors and landed elites.[23] Brahmanical territorialism continued to meet resistance and rarely travelled far into mountains, forests, deserts, swampy deltaic lowlands, or fisheries, but its hierarchical social discipline suffused all the major territories of imperial power in agrarian lowlands from Kashmir to Gujarat to Bengal and Kanya Kumari during the Medieval Warm Period, circa 900s-1200s.[24]

Those centuries of wetter monsoons enabled a broad expansion of farm acreage throughout southern Asia, boosted potential for increasing agricultural production using more intensive labour input in irrigated rice cultivation, and thereby accentuated demands for labour and resource control by local landed elites and by ambitious rulers who needed more material and labour input for more advanced, specialized, expensive armies, as well as for more elaborate investments in architecture, popular ritual, and personal consumption. These were centuries of a general increase in economic, population, and imperial expansion in Asia's southern latitudes.[25] Meanwhile, in the steppe, hotter, drier summers encouraged more southerly migration and more trade in war horses that strengthened imperial armies and economies in the south.[26]

Southerly mobility from the steppe by warriors, settlers, and merchants enhanced the value of highland valleys below the Hindu Kush, along the horse trade routes from Khorasan, where Seljuk Turks compelled an upstart *mamluk* horse-trading warrior called Mahmud to flee southeast through Indo-Persia, where he launched a new Persianate dynasty, loyal to the Caliph, in the old Buddhist commercial centre at Ghazni. His Ghaznavid Dynasty (977-1186) expanded west across Khorasan, conquered much of Persia, and repeatedly plundered southern Indo-Persia to finance wars in the north and west. Mahmud became the first Turkic Sultan, blessed by the Abbasid Caliph in Baghdad, whose spiritual authority rose among the Turk warriors who were taking control of Abbasid territory. Mahmud produced

22 Pollock, *Language*.

23 Stein, 'Integration', pp. 175-216; Heitzman, *Gifts*. See also note 24.

24 Lieberman, *Strange Parallels*, pp. 103-112.

25 Shaffer, 'Southernization', pp. 1-21; Lieberman, *Strange Parallels*.

26 As a result, 'the period after 1000 looms large in the economic historiography of early India as one of unprecedented rural expansion, which would have been barely sustainable without adequate hydraulic resources.' Chakravarti, 'Natural Resources', pp. 52-53.

foundations for Afghan imperial territory with lavish patronage for arts, sciences, and literature; he made Ghazni a cultural capital with colleges, gardens, mosques, palaces, and caravansaries; it is said that he brought whole libraries from Isfahan to Ghazni.[27]

Ghazni nevertheless remained a Buddhist monastic centre. Buddhists and Hindus enriched many market towns that Mahmud conquered; Al Biruni reported that Mahmud was passively tolerant of the non-Muslims who enriched his imperial economy, while he was fierce against enemies of orthodox Sunni Islam. The legend of his desecration of the Somanatha temple in Gujarat may derive from his reliable reputation for brutal pillage and plunder and less reliable public presentation of himself as violent iconoclast.[28] Buddhist merchants and monks continued to travel old routes across northern Indo-Persia. Ghazni seems to have followed the Bamiyan model. After Bamiyan's capture by Saffarids in 871, it remained a vibrant Buddhist centre until demolished by the Mongols in 1221; by that time, stupas and monasteries had declined from neglect, but the two colossal Buddhas remained Silk Road landmarks, described by Muslim geographers as wonders of the world, signs of god's greatness. The continued vitality of Buddhist traffic through northern Indo-Persia is indicated by Rashid al-Din's thirteenth century account of Buddhism, based on conversations with Buddhists from all over Asia who came to Iran after Hulegu's conversion to Tibetan Buddhism, following Kubilai Khan's example.[29]

Bamiyan merchants seem to have drifted slowly toward Islam over many generations, as they pursued opportunities in Muslim commercial networks with attraction by Persianate literati, artists, architects, legal scholars, and clerics (*ulema*) who produced ever more inclusive standards for Muslim territorial order, and persuaded by Sufi saints, poets, and mystics who provided entertainment, spiritual guidance, and inspiration.[30] That combination of Muslim military force, state patronage, commercialism, aesthetics, and spiritualism suffused Buddhist territory along trade routes through Indo-Persia. Sufi-inspired Islam slowly suffused the agrarian commercial Vale of Kashmir, where the Brahman Pandit elite remained entrenched after the first Muslim ruler emerged locally in 1320; and most of Kashmir was Muslim by the 1400s.[31]

27 Alikuzai, *Concise History*, pp. 67, 96-97, 121ff.

28 Thapar, *Somanatha*.

29 Elverskog, *Buddhism*.

30 Morgan, *Buddhas*.

31 Sneddon, *Understanding Kashmir*.

In the south, Turk warriors became known as Afghans, as they conquered old routes of mobility from Punjab to Bengal, carrying the same bundle of Persianate elements that shaped their own culture inside old Buddhist spaces on outskirts of imperial Brahmanism. Delhi Sultans conquered Senas in Bengal in 1204, and Islam slowly filled old Buddhist swampy agrarian frontier space in the Brahmaputra-Meghna basin, where the Turkic Sufi Shah Jalal met the Arab Moroccan Ibn Battuta, in Sylhet, in 1330,[32] indicating the spatial scale of mobility that was shaping Indo-Persia at the time. South of Bengal, on the coast through Arakan, Islam travelling through Indo-Persia met Hinduism, Buddhism, and Arab Islam sailing the Indian Ocean from Persia and Africa to Southeast Asia.[33] Meanwhile, Muslim Jats in Sindh, who converted, like Turks, as soldiers in Umayyad Arab armies, migrated to settle in more fertile northern land to become the warrior farmers of western Punjab.[34]

As patronage for Islam travelled through Indo-Persia, Muslim rulers relied on *kafirs*. Bengal Sultans relied on Brahmans, and later the Mughals fostered the multi-caste Hindu elite *bhadralok* that organized the agrarian economy and manned the imperial administration. Like early Umayyad Caliphs and the Chinggisid Mongols, Muslim Turk and Turko-Mongol warriors formed dynastic elitist layers of imperial society, variously managing their relations with lower echelons. Expansive imperial militarism thus continued to strengthen Brahmanism to form a solid territorial foundation for empires that grew and expanded from medieval to modern times.

Each subsequent empire – including the British – defeated, displaced, and subordinated the higher echelons of predecessor regimes and then sank its teeth ever deeper into local resources in territories where they depended on Hindu caste elites to sustain imperial order. The Delhi region became the central hub for imperial expansion in southern Indo-Persia, at the strategic intersection of routes running north-south and west-east along river basins supporting the richest agrarian economies and leading to ports on the coast where inland economies met the sea trade.[35] Launching their regime on the coast, the British logically expanded west and eventually relocated their capital in New Delhi.

Delhi's imperial career thus spans a medieval-to-modern millennium in which empires deepened the territorial authority of Brahmans in India, and

32 Eaton, *Rise of Islam*.

33 D'Hubert, *In the Shade*.

34 Wink, *Al-Hind*, pp. 241-242.

35 Deloche, *Transport*.

also of Theravada Buddhists in Burma and Sri Lanka. In the first millennium, monks from Sri Lanka had spread Theravada orthodoxy by sea into Burma and Thailand, and Theravada empires had arisen during medieval warm centuries when Mahayana was retreating in the face of Hindu, Muslim, and Confucian power.[36] Bamar tribes had migrated south from Yunnan into the Irrawaddy Valley from the seventh century onwards, during the southern expansion of China under the T'ang, and embraced Buddhism to build an imperial infrastructure based on agricultural expansion during the warm Bagan Period (1044-1287).

Bamar imperial territory spread down the Irrawaddy River Valley as rulers patronized the construction, it is said, of 4,446 temples, pagodas, and monasteries on the Bagan plains.[37] Trade and migration up from the Irrawaddy plains and across mountains carried Theravada into the tribal highlands where rivers flow into the Meghna and Naf Rivers, in Bengal and Arakan, then into the Bay of Bengal, on eastern frontiers of Indo-Persia, eventually forming major Theravada communities among hill people in Manipur, Nagaland, Mizoram, and the Chittagong Hill Tracts.

As we will see, Theravada Buddhist ethnic majoritarianism became predominant after Burma's independence from British India, as it also did in Sri Lanka, where the Portuguese, Dutch, and English had in turn conquered inland from the coast, where Christians and Muslims proliferated in urban markets, and where Tamil Hindu farmers maintained northern and eastern coastal territories. Throughout the agrarian lowlands, Sinhala Buddhist elites remained dominant; as soon as they acquired independence, they also mobilized the Buddhist majority in their drive to unite national territories under the banner of ancient religious tradition.[38]

Imperial modernity

All territorial transitions from empire to nation wrapped old spaces of mobility in globally expansive and locally tightening frames of physical and conceptual control. This brings us back to where we began: with the recognition that controlling human mobility is a constant, unfinished process, as are efforts to contain SARS-Covid-19; and that the process can be brutal, as shown by Black Lives Matter (and other headline news today, notably

36 Ludden, 'Decline'.

37 Lieberman, *Strange Parallels*, pp. 115-116; Aung-Thwin, *Pagan*.

38 Ludden, *India*, Ch. 8.

in India, Myanmar, and China), even as (or perhaps because) controlling mobility never fully succeeds.

There is still no history of the interaction of long-term human mobility with the global expansion of modern nation-state territorialism. Asia is a good place to begin that history; it would indicate that however out of control human mobility may seem, it is never merely random: human mobility forms spatial patterns that represent some form of organization, that is, control. Physical factors provide order in trajectories, speed, distance, and volume of mobility; and humans also organize mobility with knowledge, motivation, and power. Territorial control has typically meant controlling sites where people settle, controlling access to routes of travel, providing motives for travel and settlement, and managing technologies of travel.

Spatial histories of territoriality can thus begin by following records of human activity that organizes mobility along routes of travel, for example, by following Alexander's journey or the records of travellers on the Silk Road.[39] In a long-term view, however, spaces of mobility physically expand, and the increasing scale of mobility, over time, produces more expansive territorial controls, covering more distance, and it simultaneously generates more intensive controls over individual sites, expanding into their hinterlands, thus further increasing the volume of mobility.

Mongols dramatize this historical dynamic of spatial expansion. Three generations of Chinggis Khan's lineage organized mobility with greater speed and volume over more distance than any before. They built larger, more disciplined nomad warrior armies, with more better-fed horses (in the waning years of the Medieval Warm Period); they took control of strategic grasslands and cities, one after the other, stringing them together to form routes that enabled Marco Polo to travel from Venice to Beijing in the record time of three years (1271-1274), and then to travel safely throughout China, and finally to sail from Kublai Khan's realm to his cousin ally's Il-Khanid Persia, of course stopping on the way in South India.

The expansive space of mobility produced by Mongol territorial power also carried the Plague to Europe, where cities and towns instituted more intense local controls over mobility, which became public health measures.[40] Increasingly intense local controls over people and mobile resources also developed in and around cities that became centres of imperial power for the later Asian warrior dynasties: Ottoman, Timurid, Mughal, Safavid, Ming, and Qing.

39 Hansen, *Silk Road*.

40 Slack, 'Responses', pp. 433-453.

Building imperial territories thus entailed moving military force from place to place to control more sites, on more routes, and to position loyal subordinates in places where they could control and dispatch human and material resources up the ranks to serve higher authorities, who would live in more prestigious sites at the centre of widening networks of mobility embracing increasingly diverse populations.[41] Modern empires emerged as this old interaction of mobility and territoriality came to include global ocean travels by Europeans, enriched first by Afro-Asian trade and conquest, then by New World conquest and slavery, and then by industrial factory discipline, products, and capital accumulation.

That modern spatial transformation began in the wake of the Mongols and Plague in the fifteenth century, when consumers in Renaissance Europe bought shiploads of Asian commodities.[42] Ottomans intercepted ships to increase their own territorial control over mobile assets, driving Europeans out to sea in search of new routes to Asia, where they sought control over the spice trade at strategic Indian Ocean sites in coastal Indo-Persia. Those strategic sites became centres of capital accumulation for rulers and commercial investors who sought to control wealth and commodities moving along routes inland.[43] By 1820, the English East India Company had firm control of crucial routes of mobility across southern Indo-Persia, defining a territory that became known as India.

British India was a public-private partnership that combined military and commercial power to control routes of mobility across Indo-Persia. The limits and methods of that control defined the scale and substance of what became Indian national territory during the century between our first two global pandemics, in 1817 and 1918, which fittingly followed the Napoleonic Wars and World War One, respectively.

India's territorial limits emerged where empires met on Indo-Persia frontiers. Imperial Nepal based in Kathmandu took control of central routes through Himalayan valleys, connecting the Ganga Basin with Tibet and China. Imperial Russia conquered routes across the steppe and northern Indo-Persia. Imperial Afghans conquered from Kabul through contested mountain passages filled with local warrior territories around the Hindu Kush. Eastern, western, and southern Afghan frontiers succumbed to vague boundary making with Persia and British India in 1919 and 1921, respectively. Local Afghan warriors thus kept control of routes in and around the Hindu

41 Burbank and Cooper, *Empires*.

42 Brotton, *Renaissance Bazaar*; Mack, *Bazaar*.

43 Subrahmanyam and Bayly, 'Portfolio capitalists', pp. 401-424.

Kush, except in Kashmir, which was a loosely governed summer retreat for Mughals, bequeathed to the East India Company by the Sikhs, to become a Native State in British India under a Dogra Rajput Raja and spanning trade routes connecting Persia and Panjab, with Tibet.

It was only east and north of Calcutta that India expanded under the British beyond the eighteenth-century frontiers of Indo-Persia. After 1820, imperial armies conquered old trade routes occupied by imperial Burma[44] across mountains around Assam and across coastal Arakan to the Irrawaddy delta; opening lowlands to mobile Bengali rice-farmer settlers in the Brahmaputra, Meghna, and Barak River valleys and along coastal Arakan; opening highlands around Assam to British tea planters; and opening routes for British ambition into Tibet[45] and Yunnan.[46]

Indo-Persia's modern imperial boundaries never entailed much control over non-military travel. Old patterns of merchant and settler mobility continued, as boundaries emerged slowly, in bits and pieces, at key places along routes at points of imperial stalemate. The spatial expansion of British India followed allied ambitions of British private capital, state militarism, and aspiring landowners who paid taxes in return for imperial protection of territorial landed property rights. Border controls emerged where those ambitions were threatened. An early boundary set the pattern in 1791, when British armies pushed mobile Khasi warrior-merchant-farmers out of lowland Sylhet to make room for loyal tax-paying Bengali farmers, sketching the future northern boundary of Bangladesh at the base of the mountains of Meghalaya.[47] The imperial expansion into what became Northeast India took off from there in the 1820s.

Territorial power in British India focused squarely on the expansive intensification of profitable resource control for commercial enterprise. Resources extracted from landowners, farmers, agricultural workers, and investors in South Asia financed economic globalization in the British Empire. In 1840, eighty per cent of total British India revenue came from land taxes (sixty-five per cent) plus salt and opium monopolies (fifteen per cent). In 1860, the total was still seventy-five per cent. In 1913, other taxes decreased the land tax proportion, but by then, farmers, agricultural workers, merchants, and bankers were sending riches through thriving port cities to support the empire in many ways: workers and investors enriched British

44 Ludden, 'Cowry Country', pp. 75-100.

45 Palace, *British Empire*.

46 Cederlöf, *Founding an Empire*.

47 Ludden, 'First Boundary', pp. 1-54.

territories in the Americas, Africa, Southeast Asia, and China; farmers produced massive industrial and consumer crops for global markets and British profits; Indians bought the world's largest share of British textiles; and Indian farmers manned armies that brought British victories in the Boer Wars and World War One.

The empire invested in territorial control using all the old technologies, including broad support for Brahmans who authorize caste control over land and labour. Based in big cities and travelling among ports, British imperial investments privileged cities and elites, whose territorial wealth grew with the expansive mobility of industry in railways, ports, steamships, and global networks of capital investment. Concrete, steel, and steam engines built rigid linear trajectories of travel connecting seaports and the interior, producing a new cultural experience of space as having fixed geographic parameters, described on maps and railway and shipping schedules. Territory became state geography. State boundaries enclosed spaces of administrative authority. Surveys and maps fixed territories of property rights whose limits marked limits of empire.[48]

Empire to nation

Mobility and capital accumulation through strategic sites on railway lines – above all, the great ports of Calcutta, Bombay, and Madras; later Rangoon and Karachi – sustained British India and fed the empire's global networks. Orderly control over the empire's global mobility faced increasing disruption after 1905, when Indian nationalists began to challenge British control with mass mobilization that quickly escaped state control in strategic sites of imperial authority, most importantly, Calcutta. After 1918, managing mobility from London in all the places that defined imperial territory in its global totality became unsustainable. Like the Mughals, the British saw regional rulers become independent, in three new territories: Pakistan, Burma, and India.

The 1947 Partition was the single most violent territorial compulsion of mass mobility in human history. It displaced up to twelve million people, who instantly became refugees. It also opened an era of territorial reorganization during which imperial victors in World War Two sought to pursue global competition without provoking World War Three. That era has entailed the consolidation of territorial power by former imperial subordinates who

48 Goswami, *Producing India.*

are now officially recognized by global authorities as representatives of populations defined by immutably fixed national state territorial borders.

National ideas that mobilized popular anti-colonial passions for independence suddenly became instruments of territorial control. The identification of religious and ethnic labels with national identity and belonging and thus with qualification for citizenship became one such instrument. In 1947, that instrument organized violent mass compulsion as people labelled Hindu and Muslim stood on opposite sides of struggles for control over property in India and Pakistan. Refugees were losers in those struggles.[49] Since then, national territories around the world have annually produced more refugees and more displaced and stateless people, who now number over seventy million globally.[50]

New national borders that carved up British India produced totally new cultural borderland territories, as Hindus, Muslims, and Sikhs left places where they had mingled for centuries to live in more ethno-homogeneous regions, across the border. West Punjab became mostly Muslim for the first time, as Sikhs and Hindus left for eastern areas where they became respectively dominant in Indian Punjab and Haryana. On the southern flank of the Indo-China corridor, Sylhet lost its ancient inheritance of high caste Hindu business and intellectual elites, who left to resettled among colleagues and relatives in southern Assam.[51]

Disentangling diverse populations also left behind citizens whose ascriptive status identity suddenly became attached to another country. External borders separating nations were thus internalized inside territories carved out of old mobile spaces where all kinds of identity had travelled, settled, and mingled. People with privileged national identity status could thus brand 'others' as being less worthy and even as threats to national unity, by virtue of their identification with aliens across the border. Religious affiliation became a sign of attachment to hegemonic national norms, naturalized in cultural media, for example, in routine descriptions of India as a 'majority Hindu country,' Pakistan as 'a Muslim country,' and Sri Lanka, Myanmar, and Thailand as 'predominantly Buddhist.'

All nations in South Asia also inherited imperial status ranks that aggravate and solidify inequity.[52] Imperial elites had defined their territory as being ruled by people with their own status markers, who they deemed

49 Zolberg, 'Formation', pp. 24-38. Thanks to Dina Siddiqi for this citation.

50 United Nations, 'Global issues'.

51 Dasgupta, *Remembering Sylhet.*

52 Ludden, 'Imperial Modernity', pp. 581-601.

entitled to rule over people in subordinate ranks; and national elites continued with that strategy in pursuit of national unity, sovereignty, and progress. The most severe status deprivation in the lowest ranks produced statelessness and rightless refugees – as in the case of Rohingyas. More broadly, however, people in lower status ranks suffer impoverishment through the loss of secure entitlements to all kinds of resources,[53] from land and food to sanitation, education, healthcare, protection from violence, and freedom to move, notably, in India, for Dalits and Muslims.[54]

Old spatial dynamics of empire mapped themselves into national territory. The Republic of India revolves around Harsha's imperial territory and the capital region of southern Indo-Persia, anchored by the social power of dominant high-caste Hindus. On frontiers of India's imperial expansion, regional elites acquired independence in Nepal, Afghanistan, Pakistan, Bangladesh, and Burma, and still fight for it in Kashmir and Northeast India.[55] Old patterns of everyday mobility also continue across all the new state borders, aggravating territorial anxiety among national elites and triggering violent efforts to control and contain mobility, generating more trans-border mobility and refugee flight, further aggravating inequity.[56]

The expansion of mobility with globalization since the 1980s has again logically fostered more aggressive efforts to intensify control over resources inside national territory, now for the benefit of globe-trotting capital investors. The identification of religion with national unity has become a rallying cry for mobilizing national support from demographic majorities to deprive minorities of entitlements to resources. The violence compelling Rohingyas to flee Arakan is a telling reiteration of Partition, which has now been followed by military attacks on Rakhine Buddhists, to expose the Burmese military's determination to control spaces of mobility in Arakan and to control profits from Chinese investments in the new Silk Road.[57] The mobility of capital and of imperial territorialism continue to work together in the world of national states.

The Black Lives Matter movement compels our attention to the violent legacy of slavery and other coercive bodily social controls that underlie territorial controls feeding all kinds of inequity. On Christmas Day in

53 Sen, *Poverty*.
54 Rahman, *Denial*.
55 Webb, *Separatist Violence*.
56 Uddin and Chowdhory, *Deterritorialised Identity*.
57 Ibrahim, *Rohingyas*.

1968, Dalit agricultural workers who had formed a union to seek better working conditions and higher wages hoisted red flags to celebrate their solidarity in the Tamil Nadu village of Kilvenmani, where high-caste landlords then burned their homes to the ground and burned forty-four Dalit men, women and children to death. Many hundreds of violent attacks on Dalits have occurred in India since then, often attending the achievement of some social advancement. Violence classified as 'Hindu-Muslim riots' in India is also typically launched by violence against Muslims in similar efforts to suppress threats to social power.[58] Violence against women and minorities in all the countries of South Asia similarly remaps the power dynamics of imperial territorialism into the global discourse of national development. The slavery that mobilized *mamluks* for the Delhi Sultans' military and starving peasants for building Dutch Batavia early irrigation works in British India may have disappeared into modern labour markets where poverty and debt still do the work of coercion, but southern Indo-Persia remained distinguished by Hindu caste power relations that propelled the intensification of agriculture from medieval times and still lurk in violent inequity in India's majoritarian globalization.[59]

References

Aberth, John. *Plagues in World History.* Lanham: Rowman and Littlefield, 2011.

Alikuzai, Hamid Wahed. *A Concise History of Afghanistan in 25 Volumes: Volume 14.* Victoria BC: Trafford Publishing, 2013.

Aung-Thwin, Michael. *Pagan: The Origins of Modern Burma.* Honolulu: University of Hawaii Press, 1985.

Brotton, Jerry. *Renaissance Bazaar: From the Silk Road to Michelangelo.* Oxford: Oxford University Press, 2002.

Burbank, Jane, and Frederick Cooper. *Empires in World History: Power and the Politics of Difference.* Princeton, NJ: Princeton University Press, 2010.

Cederlöf, Gunnel. *Founding an Empire on India's North Eastern Frontiers, 1790-1840: Climate, Commerce, Polity.* New Delhi: Oxford University Press, 2014.

Chakravarti, Ranabir. 'Natural Resources and Human Settlements: Perceiving the Environment in India'. In *The World in the Year 1000*, edited by J. Heitzman and W. Schenkluhn. Lanham. MD: University Press of America, 2004.

58 Mitra and Ray, 'Implications', p. 719-765.

59 Ludden, 'Imperial Modernity'.

d'Hubert, Thibaut. *In the Shade of the Golden Palace: Alaol and Middle Bengali Poetics in Arakan*. Oxford: Oxford University Press, 2018.

Dasgupta, Anindita. *Remembering Sylhet: Hindu and Muslim Voices from a Nearly Forgotten Story of India's Partition.* New Delhi: Manohar, 2014.

de Certeau, Michel. *The Practice of Everyday Life*. Berkeley, CA: University of California Press, 1984.

Deloche, Jean. *Transport and Communications in India prior to Steam Locomotion.* New Delhi: Oxford University Press, 1993.

Digby, Simon. *War-Horse and Elephant in the Delhi Sultanate: A Study of Military Supplies.* Oxford: Orient Monographs, 1971.

Eaton, Richard M. *The Rise of Islam and the Bengal Frontier, 1204–1760*. Berkeley and Los Angeles, CA: University of California Press, 1994.

Elverskog, Johan. *Buddhism and Islam on the Silk Road.* Philadelphia, PA: University of Pennsylvania Press, 2010.

Goswami, Manu. *Producing India: From Colonial Economy to National Space.* Chicago, IL: University of Chicago Press, 2004.

Hansen, Valeria. *The Silk Road: A New History.* New York, NY: Oxford University Press, 2012.

Heitzman, James. *Gifts of Power: Lordship in an Early Indian State.* New Delhi: Oxford University Press, 1997.

Ibrahim, Azeem. *The Rohingyas: Inside Myanmar's Hidden Genocide.* London: C. Hurst and Company (Publishers) Limited, 2016.

Lieberman, Victor B. *Strange Parallels: Southeast Asia in Global Context, c. 800–1830.* Cambridge: Cambridge University Press, 2003.

Ludden, David. 'The First Boundary of Bangladesh on Sylhet's Northern Frontiers'. *Journal of the Asiatic Society of Bangladesh* 48, no. 1 (June 2003): pp. 1–54.

Ludden, David. 'Maps in the Mind and the Mobility of Asia'. *Journal of Asian Studies* 62, no. 4 (2003): pp. 1057–1078.

Ludden, David. 'Imperial Modernity: History and Global Inequity in Rising Asia'. *Third World Quarterly* 33, no. 4 (2012): pp. 581–601.

Ludden, David. *India and South Asia: A Short History.* 2nd ed. London: OneWorld, 2014.

Ludden, David. 'The Decline of Buddhism, Revisited: The Rise of Hindu Territorial Hegemony, c.950–1250'. In *Clio and Her Descendants: Essays for Kesavan Veluthat,* edited by Manu V. Mahadevan, pp. 132–162. New Delhi: Primus Books, 2018.

Ludden, David. 'Cowry Country: Mobile Space and Imperial Territory'. In *Asia Inside Out. Volume III, Itinerant People,* edited by E. Tagliacozza, H. F. Siu, and P. Perdue, pp. 75–100. Cambridge, MA: Harvard University Press, 2019.

Mack, Rosamond E. *Bazaar to Piazza: Islamic Trade and Italian Art, 1300–1600.* Berkeley, CA: University of California Press, 2002.

Manjapra, Kris. *Colonialism in Global Perspective.* Cambridge: Cambridge University Press, 2020.

Mitra, Anirban, and Debraj Ray. 'Implications of an Economic Theory of Conflict: Hindu-Muslim Violence in India'. *Journal of Political Economy* 122, no. 4 (2014): pp. 719–765.

Morgan, Llewelyn. *The Buddhas of Bamiyan.* Cambridge, MA: Harvard University Press, 2012.

Neelis, Jason Emmanuel. *Early Buddhist Transmission and Trade Networks: Mobility and Exchange Within and Beyond the Northwestern Borderlands of South Asia.* Leiden: Brill, 2010.

Palace, Wendy. *The British Empire and Tibet 1900–1922.* London: Routledge, 2004.

Pollock, Sheldon. *The Language of the Gods in the World of Men: Sanskrit, Culture, and Power in Premodern India.* Berkeley and Los Angeles, CA: University of California Press, 2007.

Rahman, Abdur. *Denial and Deprivation: Indian Muslims after the Sachar Committee and Rangnath Mishra Commission Reports.* London: Taylor & Francis Group, 2018.

Reliefweb. 'Conflicts without borders', https://reliefweb.int/sites/reliefweb.int/files/resources/4307B727E9530039852576B10054EA19-map.pdf, accessed 23 February 2021.

Rienjang, Wannaporn, and Peter Stewart. *The Geography of Gandharan Art: Proceedings of the Second International Workshop of the Gandhāra Connections Project.* Oxford: Archaeopress, 2018.

Roy, Kaushik. *Military Manpower, Armies and Warfare in South Asia.* London: Routledge, 2013.

Sacks, Milton. 'The Strategy of Communism in Southeast Asia'. *Pacific Affairs* 23, no. 3 (1950): pp. 227–247.

Schwartzberg, Joseph E., ed. *A Historical Atlas of South Asia.* New York, NY: Oxford University Press, 1992.

Sen, Amartya. *Poverty and Famines: An Essay on Entitlement and Deprivation.* Oxford: Oxford University Press, 1981.

Sen, Tansen. 'The Travel Records of Chinese Pilgrims Faxian, Xuanzang, and Yijing: Sources for Cross-Cultural Encounters between Ancient China and Ancient India'. *Education about Asia* 11, no. 3 (2006): pp. 24–33.

Shaffer, Lynda. 'Southernization'. *Journal of World History* 5, no. 1 (1994): pp. 1–21.

Slack, Paul. 'Responses to Plague in Early Modern Europe: The Implications of Public Health'. *Social Research* 55, no. 3 (1988): pp. 433–453.

Sneddon, Christopher. *Understanding Kashmir and Kashmiris.* London: Hurst and Company, 2015.

Stein, Burton. 'Integration of the Agrarian System of South India'. In *Land Control and Social Structure in Indian History,* edited by Robert Eric Frykenberg, pp.175–216. Madison, WI: University of Wisconsin Press, 1969.

Subrahmanyam, Sanjay, and Christopher A. Bayly. 'Portfolio capitalists and the *political* economy of early modern India'. *The Indian Economic and Social History Review* 25, no. 4 (1988): pp. 401–424.

Thapar, Romila. *Somanatha: The Many Voices of a History*. London: Verso, 2005.

Uddin, Nasir, and Nasreen Chowdhory, eds. *Deterritorialised Identity and Trans-border Movement in South Asia*. Singapore: Springer, 2019.

United Nations. 'Global issues, refugees', accessed 4 August 2021, https://www.un.org/en/global-issues/refugees.

Verardi, Giovanni. *Hardships and Downfall of Buddhism in India*. New Delhi: Manohar, 2011.

Webb, Matthew J. *Separatist Violence in South Asia: A Comparative Study*. London: Taylor and Francis, 2016.

Wink, André. *Al-Hind, The Making of the Indo-Islamic World: Slave Kings and the Islamic Conquest, 11th-13th Centuries*, vol. 2. Boston: Brill, 2002.

Zolberg, Aristide R. 'The Formation of New States as a Refugee-Generating Process' in 'The Global Refugee Problem: U. S. and World Response', special issue, *The Annals of the American Academy of Political and Social Science* 467 (May 1983): pp. 24–38.

About the author

David Ludden, Professor of History, New York University, USA. His research focuses on agrarian Bangladesh and South and Northeast India. His monographs include *India and South Asia* (2013), *An Agrarian History of South Asia* (2011), and among his edited volumes are *Making India Hindu* (2005) and *Reading Subaltern Studies* (2002).

A Long View

3 The Road Towards *All under Heaven* Cosmology

The Bazi Basin Society in West Yunnan

Jianxiong Ma

Abstract

This chapter explores the historical dynamics of 'basin societies', which colluded with the state to maintain transportation systems between Yunnan and Burma. Some social networks overlapped with transportation routes as these routes had guaranteed the successful delivery of tribute from exterior chieftains to the capital since the early Ming – and thereby confirmed the 'All Under Heaven' state cosmology. The Zhaozhou Bazi basin has been regarded as sustaining a geopolitical network of states. Pushed by the local elite, communities reorganized their common property to address changes in state policies and extended their networks to search for alternative routes. Thus, dynamic local agency actively reconstructed basin societies and affected centre-periphery relationships in southwest China.

Keywords: Bazi society, All Under Heaven cosmology, common property, Yunnan province, Southwest China, social network

Southwest China and the highlands of southeast Asia are made up of a landscape of high mountains, particularly in Western Yunnan, where there is a region of ragged mountains and deep river gorges of the Upper Yangtze, the Mekong, the Red, the Salween, and their numerous branches.[1] In this region, the basins had been regarded as a sustaining support of a

1 The author acknowledges a grant supported by the Research Grants Council (Hong Kong), Project No. 16655916.

Cederlöf, Gunnel, and Willem van Schendel (eds), *Flows and Frictions in Trans-Himalayan Spaces: Histories of Networking and Border Crossing*. Taylor & Francis Group 2022
DOI: 10.5117/9789463724371_CH03

geopolitical network by states. In this chapter, the author discusses the process of communal reconstruction from the fourteenth century to the eighteenth century that was based on an interconnection of basins in the mountainous landscape. The social reconstruction was bound to the flows of goods that cooperated with the flows of relations exchanged between the basins. Because these flows were based on social, political, and commercial support that was serviced by the basin societies as links in the transportation network, the transportation network has been the frame of local ethnic relations and cultural connections. The author regards the system of basins as an open space toward Southwest Asia and Tibet. The construction of basin societies represents a form of state presence in a mountain area on the borderland. This is explained by the responsibility that the communities and the imperial officials in one basin had to manage the road across the mountains into the next basin which, thereby, developed local agency as a response to the demands of the state institutions. In this process, the local elite – religious masters, scholars, and village gentries – performed important roles as local actors. Due to state-policy reforms, the land tax and its corvée for transport services had increased the burdens of local communities after the 1570s. In response, pushed by the local elite, communities reorganized their common property to deal with changes in state policies and extended their networks to search for alternative opportunities over the transportation routes. In addition, the basins had been a growing social space with the extension of state power and state cosmology, and the basin society was integrated into a macro scale of *All under Heaven* cosmology, along with the social reconstruction and state institutional reforms. The *All under Heaven* cosmology presents a symbolic universe in traditional China, which emphasizes cultural continuity and aims to 'use Chinese ways to transform the barbarians' as the core of Confucianism symbolism.[2]

The basin areas are treasured by local people in Southwest China, especially in Yunnan and Guizhou provinces, and they call the basins the Bazi. Among these Bazi basins,[3] the big basins are the most important

2 Dikötter, *Discourse*, pp. 2-3.

3 This chapter is based on outcomes of a long-term cooperative research project on Bazi society in Southwest China between the author and colleagues at Dali University in Yunnan since 2009. In Yunnan and Guizhou provinces in Southwest China, most of the area is more than 1,000 metres above sea level. In northwest Yunnan, the highest mountain peak is 6,740 metres above sea level, but the valley of the Red River on the border between Yunnan and Vietnam drops sharply down to 76.4 metres above sea level. In this realm of numerous ragged mountains in Yunnan, only about 6.52 per cent of its territory consists of small flat basins and relatively flat land surrounded by mountains. Therefore, more than 93 per cent of the geographic terrain

political and cultural centres and economic bases in Yunnan. The basins are linked by transportation routes and are concentrations of cities and towns and should therefore be seen as the geographic framework of junctions of the transportation system, the hubs of economic connections between cities and towns in the mountain areas of Yunnan or Southwest China. This interpretation helps to go beyond the limitation of theoretical dichotomy of highland and lowland societies in this region suggested by James C. Scott, and to provide a different angle from which to study the relationship between state and society in the mountainous areas.[4] This chapter focuses on the development of the historical unit of social reconstruction since the 1380s by taking the Zhaozhou basin in Yunnan as a case to analyse. When we discuss the well-used concept of Bazi in the Han Chinese language in Yunnan today,[5] to describe a sociopolitical system centred on a flatland surrounded by mountains.[6] All those Bazi systems were linked not only by transportation systems but also by flows of goods and markets, as well as by state power. This chapter mainly focuses on the historical transmission of the Bazi basin societies, regarding the basins as the supporting positions of this system in a ragged mountainous realm in history. In order to simplify the reading of this chapter, the word 'basin' has been adopted in describing this historical construction of Bazi basin society.

Although the chapter emphasises the Bazi system's organisation and interconnectedness across a large region, it is important to remember that this system rested on social, political, and commercial support in the basins. During periods of conflict and unrest, powerful people in the basins could block flows of people, and goods and movements in the region would slow down and find alternative routes. Such changes can be seen across space and time in the Shan-Dai region, as Gunnel Cederlöf shows in this volume. The internal marriage network, which regulated the nuptials of sons and daughters of ruling families, provided a mechanism for political interference and manipulation. Alliances formed across basins, which both moderated potential conflicts and worked as a means by which two sawbwas could challenge a third. This shows how flows across the

of Yunnan comprises ragged mountains and river gorges. Among these Bazi basins, there are forty-two that we can define as the 'big Bazi', their sizes varying between 50 and 100 square kilometres. See Shaoyu and Yongsen, *Study*, p. 2, and Daniels and Ma, *Transformation*.

4 Scott, *Art*, pp. 41-63.

5 Note: The term was 'Ba' borrowed from the Tai-Dai language in history and well adopted in different languages to describe the flatland surrounded by mountains in southwest China.

6 Fuyi, *Letters*, p. 125.

Corridor, explained by Willem van Schendel and Cederlöf in this volume, found unexpected routes due to the dynamism of the interrelationships between the Bazi basins.[7]

Scholars have tried to explain the sociohistorical features of the Southwest frontier in different ways in the past several decades. As Giersch points out, the concept of 'middle ground' could be used to explain the borderland of southwest Yunnan as an area of flows of social actors and flows of cultural, economic exchanges, as well as the flow of boundaries.[8] In the frontier between Yunnan and Southeast Asia, long-distance trade in the commercial circulation industry also pushed the social change of local society; the fluidity and networks of circulations therefore also established the basis for the transformation to modernity in this region.[9] Besides this discussion, the author argues that the way in which the social agency of local communities was reconstructed is the basis used to support and maintain these flows in the long-term historical process in these borderlands. Another scholar, Hill, points out that, in northern Thailand, the Yunnanese were regarded as strangers and called 'the haw' by indigenous people to some degree. However, the Yunnanese merchants performed the role of middlemen between indigenous groups and other businessmen in this borderland, so the Thai government defined them as a group of natives.[10] This observation about the middleman role is significant for our explanation for the relationship between the basins and the mountain communities, and this role could be a common phenomenon throughout the region. But if we study the transportation networks supported by the basins, it is a crucial point that the basin societies are open for their discursive and extensive stretch toward different dimensions linked and bound by other basins and woven by their agents.[11] This social feature could help us to better understand the strong local agency of continuing communal reconstruction, which is not only visible in the political dynamic of ethnic identities, but also in their self-managed communal agency responding to state powers. In this way, considering it as a system of social infrastructure, studying the basin society is a way to explain the dialectic process of discursive flows of goods and cultural meanings and the integration of local communal unities in a

7 Ma, 'Constructing'. Cederlöf, 'Tracking', Chapter 4 in this volume, and van Schendel and Cederlöf, 'Flows', Chapter 1 in this volume.

8 Giersch, *Asian Borderlands*, p. 7.

9 Giersch, 'Cotton', pp. 37-61.

10 Maxwell Hill, *Merchants*, p. 9.

11 Ma and Ma, 'Mule Caravans', pp. 237-257.

bigger picture of state extensions since the fourteenth century and local transformation in the same process.

Since the 1950s, the newly developed academic paradigm on ethnic minorities in Southwest China has paid ample attention to the history of ethnic nationalities or the origin of Non-Han minorities in the context of the establishment of minority autonomous regions in China. In this political atmosphere, Fang Guoyu and You Zhong have tried to include the study of 'frontier society' in history into the study of ethnic history for several decades.[12] Meanwhile, John Herman and Laura Hostetler tried to adopt Michael Hechter's term 'internal colonialism', which refers to territorial claims within nation-states or integrated land empires, to discuss social reforms in this region, regarding Southwest China as part of a colonial project launched by the courts of the Ming and Qing empires.[13] Fang Guoyu and You Zhong ignored the history of the social and cultural changes of local communities and their relations with the state; John Herman and Laura Hostetler fail to review the diverse ecological environment and parties participating in state politics in the geographic landscape of the Southwest frontier of China. Through the study of basin society, we could realize that social transformation for previous local kingdoms to the eighteenth century, was not internal colonialism concerning the social reconstruction of local communities along with basins, and it was not a direct result of the civilizing project of state power.[14] The local elite performed crucial roles in negotiating with the state in the long-term course to mobilize local communities and to search for opportunities to respond to the changes in state policies. They were quite skilful at reconstructing local agency based on the development of native Confucian schools and religious reform to build up a local space for communication networks through the reform of basin societies.[15]

In brief, we could not simply regard a diverse region of mountains and basins, their supported transportation routes and branches, the different administrative styles of chieftain territories and official counties, all of this political geographic diversity in a historical dynamic, as a homogenized Southwest China frontier. Some recent studies also merely regard local communities as a simple 'passage' or 'corridor' for passing travellers, and these travellers represent the view of the central plain again, but these recent studies ignore the local social agency of communities in the reform

12 Guoyu, *Research*, p. 4. Zhong, *Collected Works*, p. 546.
13 Herman, *Amid the Clouds*, p. 15. Hostetler, *Qing*, p. 30. Hechter, *Internal Colonialism*.
14 Elvin, *Retreat*, pp. 216-272; Herman, *Amid the Clouds*, p. 15.
15 Ma, 'Local Communities', pp. 43-75.

of this frontier.[16] The perspective of basin study benefits us to think about two questions: firstly, the basin society was a historical process around the construction of local communities to deal with the pressure of state power by responding to policies like corvée and tax, the pressure of providing a transportation service, wars, the flow of goods, and so on. Secondly, the basin society was involved in continuing competition and conflicts in searching for different political and social possibilities in the whole process, and this dynamic also pushed the reconstruction of different interest groups and common identities. In this sense, the perspective of the basin study is not about a comparison of the demarcation between highland and lowland societies, as this point of view never put local agency into a network of transportation in the state political cosmology and political geography. Edmund Leach pointed out that ethnic identities and political systems in the mountain area in this region were very dynamic,[17] but we could go beyond these oscillating models to check where and how the state and the elite worked together in shaping local agency for change, social fluidity, and continuity. Thus, by taking the case of the Zhaozhou basin, we can see that, pushed by Confucian scholars, village gentries, and governmental officials, local residents gradually reformed their social landscape into a mythical and cosmological space based on the *Fengshui* geomancy theory and practice for their re-imagination of social relations with the state.

Historical connections of the basin societies in Southwest China

Between basins, both the state and local communities have mutually cooperated in the reconstruction and maintenance of the transportation system as well as the reformation of communities themselves since the 1380s. From Yunnan to the Shan highlands in Northern Burma, transportation meant travelling from one basin to another across mountains and rivers. However, we can find that systems created by local markets in the basins, agricultural irrigation canals, and local temple connections were effectively supported by local communities both in basins and nearby mountains to function as a social mechanism for local integration. Additionally, higher level, cross-region networks existed, such as the pilgrimage networks or trade fairs made possible by mule caravans.[18]

16 Shi, *Tibetan and Yi Corridor.*
17 Leach, *Political Systems*, pp. 9 and 282.
18 Ma, 'State', pp. 1-19.

The social mechanism of basin societies in the mountain area was different from the system that could be regarded as the gradual development of a parallel system of towns and cities with local agriculture and markets, from lower-level to high-level economic and political centres, as William Skinner points out. According to Skinner, Southwest China should be regarded as a part of the Chinese domestic market, and local markets were also constructed on the basis of the local administrative system, like other regions in China.[19] However, we found that the real situation is much more complex than Skinner suggested. On one hand, a regional market system must be shaped by the geographic significance of the transportation accessibility between basins. On the other hand, local marketable products such as minerals, medicinal herbs, animals, as well as the special need for regular long-distance connections between Yunnan, Burma, and Tibet and numerous chieftain tributaries, are all factors that have seriously limited the promotion of local markets based on local agriculture. In sum, the transportation system among basins was not a system simply based on local agriculture and the needs of local peasants, but a system based on state politics as well as regional transportation needs for long-distance exchange in politics (like tributary journeys), and state strategic planning (like copper transportation in the Qing dynasty). Counter to this, the transportation system between the Yangtze river ports, or Dongting Lake and the Irrawaddy river, run by animal caravans was substantially shaped by the level of local participation in regional markets and the roles performed by local communities, to form a very large network in general. That means, as a geographic position on the frontier, a local basin society had to work with others in an overall system, so that they could share a common marketable value and political resources along with transportation routes between China and Southeast Asia.

In this way, the flows of goods need to cooperate with the flows of relations exchanged between basins, and it must be an open space towards Southwest Asia and Tibet, but we can observe that these flows were based on social, political, and commercial support serviced by basins as ties of the transportation network. Additionally, this transportation network has also been the frame of local ethnic relations and cultural connections. This is not an isolated end or corner of the Chinese domestic market suggested by Skinner, but multiple levels of social networks woven by local communities, state power, and commercial flows. In the case of the Zhaozhou basin, as a transportation position on the main passage from Kunming to Burma or Tibet, we can see that the holistic scope of a transportation system for

19 Skinner, 'Marketing', pp. 3-43.

the Chinese empire became a serious shaping power in reorganizing local relations among communities and the cosmology as a cultural landscape.

The Zhaozhou Bazi: A basin society as a sustaining position in the transportation system

Today's Fengyi township in the Dali prefecture of Yunnan was previously Zhaozhou subprefecture before 1912, when the title changed to Fengyi County. The territory of this county then included the main area to the south of Erhai Lake. Because of economic development and the construction of the Burma Road in 1938, as well as the shift of the centre of the administrative district after 1958, Fengyi lost its position as the county seat and became a township-level administrative district under Dali County. According to historical records, the total population of Fengyi County was 43,000 in 1919.[20] The Fengyi township remains a transportation crossroads in western Yunnan.

In this section, based on local inscriptions, local gazetteers, and official archives, the author takes the Zhaozhou basin as a geographic unit for analysis.[21] The Zhaozhou basin is not organized by orders of ethnic groups, even though there are many official identified ethnic nationalities such as the Bai (the Minjia), the Han, the Hui Muslim, and the Yi (the Laluo) today. The basic units of social activities have long been villages and neighbourhoods in the towns. However, competition, conflicts, and social adjustment also performed important roles in the cooperation of these village communities around the management of temple-based common property, and this cooperative mechanism helped to build up a local identity. In addition, the case of Zhaozhou also provides a historical perspective on a local scale with which to check change and continuity based on integrated local communities and to study the process of transformation from a community controlled by famous noble families based on their family Buddhist monasteries in the Dali kingdom, to become a bureaucratic subprefecture after the Ming governance was established.[22]

20 Yang, *Fengyi Gazetteer*, p. 93.

21 Note: In 2001, the author and a local scholar, Mr. Ma Cunzhao, initiated a project to collect data on caravan muleteer oral history in west Yunnan and inscriptions from villages in Zhaozhou. This project was supported by Mr. Philip Chien Yi-bang. In 2008, the collection of muleteer oral histories was published in Chinese in Kunming, and in 2013, the collection of village inscriptions was published in Hong Kong.

22 Ni, *Chronicle*, pp. 241-260.

Map 3.1 The Shan-Dai region on the frontiers of Burma and China

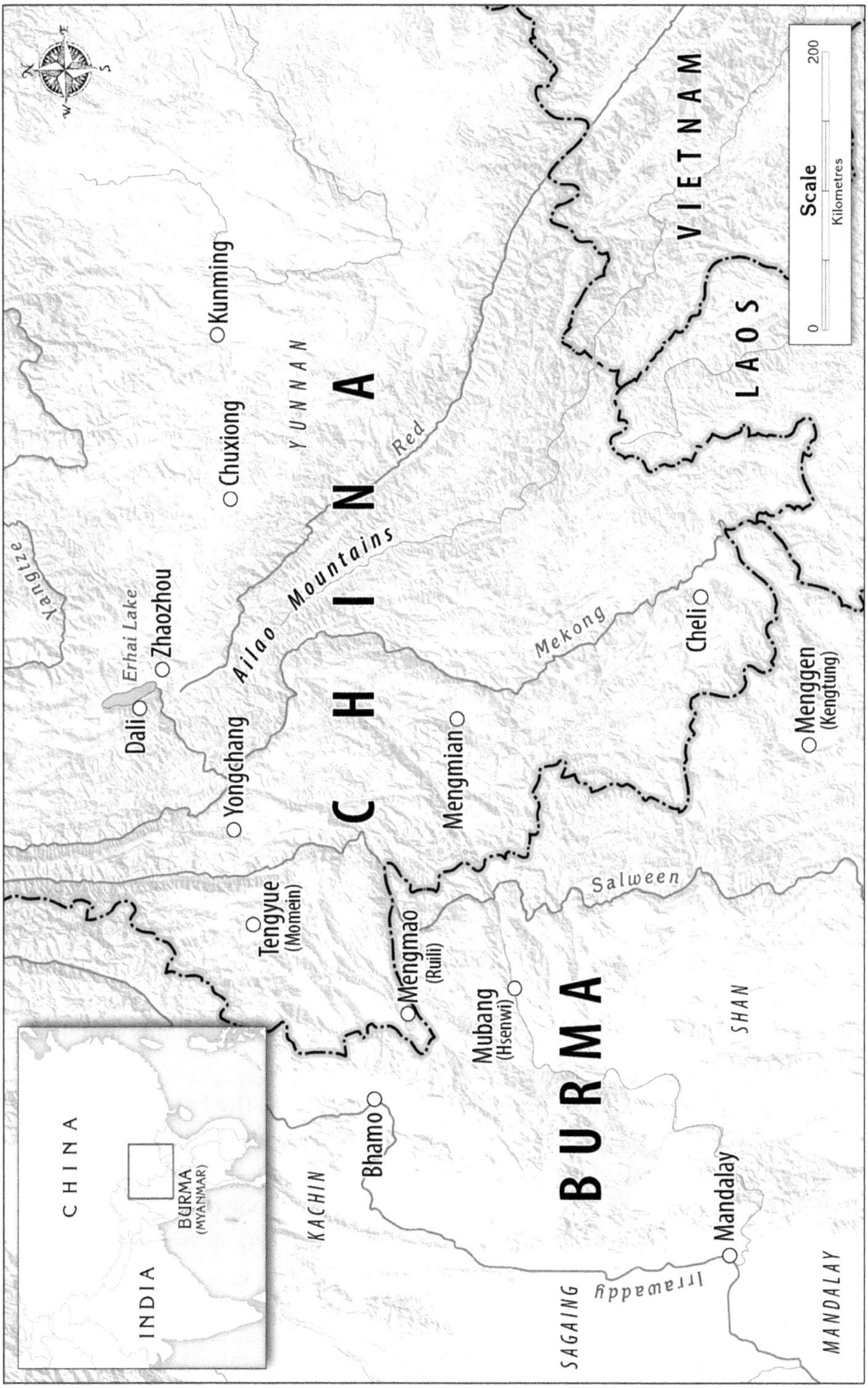

Drawn by Laurie Whiddon

Social integration and village circles of the basin society

As the political centre of the previous Nanzhao (783-902) and Dali (937-1253) kingdoms in Yunnan, in order to control the main transportation route in western Yunnan, the Ming government (Ming dynasty, 1368-1644) established the military garrison system in Yunnan in 1382. This was designed to settle the troops based on military training and part-time work on state-owned tracts, known as the state farms, with the household registration of the military households (*junhu*, the Han later). The military garrison system safeguarded the main transportation routes from the provincial capital Kunming to Burma.[23] Meanwhile, the native commoners were registered as the civilian households (*minjia*) under the bureaucratic prefecture, subprefecture, and county governments. Therefore, two different household registration categories followed: the civilians under the prefecture and county bureaucracy, and the military households under the military garrisons. In this double system civilians paid their tax and corvée to the county or subprefecture government, while the military households paid theirs to the military officials, until the system was abolished in the 1680s. After Zhaozhou subprefecture was established in 1382, all civilian households were registered and organized into fifteen *lijia* tithing units. This was a community and self-monitoring system, to designate every 110 households as a community (li), whose 10 most affluent households were appointed as community heads in annual rotation. The remaining 100 households were divided into 10 clusters, each with 10 neighbouring households constituting a tithing (*jia*), for the purpose of tax collection and corvée services.[24] In 1713, in the Qing dynasty (1644-1911), the collection of grain tax and corvée service based on the categories of household registrations changed to become a new system based on fixed ratios from cultivated land and poll tax, which was paid in silver. It was known as the reform of spread roll tax into acres (tangding rudi); the revenue collection was eventually simplified into land-and-labour-service tax.[25]

For centuries of social reforms, after the 1720s, the Qing government was able to extend its direct political control towards more remote areas in the borderland with Burma, and after more southern Yunnan areas were integrated into the county system through the abolition of native chieftains, the functional position of Zhaozhou in the transportation system changed

23 Zhao, 'Salt', pp. 19-42.

24 Hucker, *Dictionary*, p. 303.

25 Ni, *Chronicle*, p. 560.

again.[26] From the 1380s to the 1640s, Zhaozhou subprefecture gradually changed from being a frontier region to become a well-developed 'interior county' in terms of direct control of bureaucratic prefectures and jurisdiction in local affairs.

The social integration meant that, on one hand, local communities could easily organize themselves as a cooperative to adopt and respond to the coming of the Ming and the Qing states; on the other hand, this social integration also meant that a group of local elite developed who were leaders of local communities supported by the civilians to be their representatives to manage their public affairs and common properties for the common good. This leading elite also performed the roles of local representatives to work with officials and negotiate with representatives of state power for their local interests, or to work for the state to maintain their own communities. This layer of social elite comprised crucial actors who were also the middlemen for social reform and cultural reconstruction in the whole process of integration to build up a basin society.[27] Following the political reform of 'integrating military garrisons into counties' carried out by the Qing government in 1687, the separation of taxes and corvées became blurred, but the corvées of transportation services were still onerous for communities along the road.[28] The social integration of household-registration-based identities gradually transformed into ethnic identities in local contexts, mainly becoming the civilian Minjia (the Bai today) and the Han, but the social and cultural separation between basin communities and mountain residents, known as the Laluo, was still very significant, although the integration of basin communities remained ongoing in parallel.[29]

Since the 1640s, the space of the basin has been well integrated into a system of circles of transportation routes, towns, villages, and temples centred around Zhaozhou city, the seat of subprefecture government. Based on Zhaozhou city, the four city gates were defined as the four directions of villages in this basin and the surrounding mountains. All villages in the basin have been identified as having four classifications based on the four-city-gate-oriented communities, thus, all the villages were organized around Zhaozhou city, the centre of local cultural practice. For instance, all festival parades were to be organized in line with the categories of the city-gate base and village streets, and funeral parades also needed to follow

26 Ma, 'Salt', pp. 1637-1669.

27 Ma, 'Zhaozhou', pp. 131-155.

28 Ni, *Chronicle*, p. 539.

29 Ma, 'Local Communities', pp. 43-75.

this order.[30] The influence of the city could extend toward all villages in the basin and to the communities in the mountains through the transportation route or sub-routes. The system of circles, based on 'the county city – villages in basin – communities in surrounding mountains', therefore became a hierarchical space and formed the political order of the social life of different local residents in the area.

Different residential communities became villages as this system gradually became established. There were villages of transportation stations set up by the local police office, the postal relay stations, and local military companies under the subprefecture government and provincial military commander along with the roads. Some military stations designed to safeguard the transportation road set up in the time of the Qing government developed into new villages at important localities by the roadside in mountain areas. All these types of villages were communities functioning as basic service stations to protect and maintain the main transportation system in Zhaozhou.[31] However, most villages were agricultural communities locally organized by the irrigation system, like the streams, irrigation channels, and small reservoirs, but all this agricultural infrastructure became based around the village-run temples after the early years of the Qing period when the military garrisons were demolished. In sum, we can see that market towns and transportation villages, as well as agriculture-based villages and their common temples, were all linked to the county seat city based on their geographic location to the direction of the city gates.[32] Or, put another way, the county city became the centre of the small cosmology of a basin society, while this small cosmology also functioned as a support position in the bigger cosmology of the Chinese empire, when the transportation system bound together the frontier chieftains and foreign satellite kingdoms, such as Burma.

Meanwhile, another style of social cooperation developed in mountain areas. After the 1710s, the *lijia* system did not function well in basins and mountains due to the reform of revenue policies, but the cattle cooperation system performed an important role in the social organization of the mountain communities. The fundamental function of a cattle cooperation was to graze the animals. Residents in mountain communities lived scattered on the mountainsides. In order to organize their labour effectively, people

30 Cheng, *Qianlong Gazetteer*, p. 32.

31 Ibid, p. 33.

32 Wang, 'Draft', p. 364.

organized themselves into cooperative teams for herding their cattle. For instance, some neighbouring families, kinship relatives, or friends could gather their cattle together for herding, and they became a cooperative group of the *Cong* cooperation, which means cooperative shares for a common purpose in local languages. Normally in this cooperation, a group would make a board out of an ox horn. The names of all participant families were listed on the board along with the number of their cattle; then the head of this group could assign one labourer to herd all the cattle for the participant families or hire a labourer to do the same. All participant families had to pay a certain sum of money based on their number of cattle, or they could also take turns to herd the cattle. For example, to put one ox into the cooperation, the ox owner should take one day's responsibility to herd all the cattle for the group; then, others would take their turns. Every morning, the person whose turn it was to herd the cattle for the group had to blow the ox horn on the village road to inform all the families that they needed to drive their animals together. In the evening when the herder returned, he blew the horn again to ask the cattle owners to collect their animals. There was another style of organizing the herding: in the autumn and winter, all families put their farm cattle together as a group, and the selected head of this group arranged a herder to graze the cattle in a mountain valley until the following spring when the families needed their cattle for ploughing. In this way, the organized group constituted a cattle cooperation. This group was organized through herding, but it could also cooperate economically. All families pooled a certain amount of cash together every month, then each of them could use the money in turns, month by month. This style of cooperation was normally written as *Cong* too.[33] Therefore, the Cong cooperative could be labour based or cash-based, but it could also become a militia system when the mountain communities sometimes became militarized.[34]

In general, if we take the cattle cooperation systems as the exterior circle of basin society, we can find that communities in the mountains and basin were managed in different ways, and their connections with county officials varied. From the *Cong* groups in the mountains to the villages in the basin and county city, social space was constructed based on the political hierarchy under state influence along with the transportation system; boundaries of communities and identities were therefore also established along with these routes.

33 Lu, *Gazetteer.*

34 Liu, *Tianqi Gazetteer*, pp. 272-274.

The change of Vajrayana Buddhist masters and the development of village temples

During the Yuan dynasty in Yunnan (1253-1382), the masters of Vajrayana Buddhism such as the Dongs and the Zhaos remained and their social status was not destroyed in western Yunnan as the descendants of the previous kings of the Dali kingdom still controlled western Yunnan.[35] Once the Ming army occupied the Dali area, they immediately destroyed the institutions of the economics-based religious masters and thus their political power. Considering their local political and religious influence, the Ming government encouraged the previous Dali nobles to become officials to serve the Ming government; subsequently, more and more educated elite from the Zhao and Dong families joined the Ming government. The period from the 1490s to the 1520s was a significant time when local society changed quite dramatically. The changes in social relations and cultural life occurred because the Ming government strictly suppressed the religious masters of Vajrayana Buddhism and the reform of local customs. For instance, the Ming government set up two special official departments to manage the affairs of Vajrayana Buddhist monks and native shamanism and rituals specialists in Dali prefecture and some western Yunnan counties. In 1507, Yunnan provincial government demolished these two departments.[36]

A group of villages and subprefecture city-based Confucian literati and retired officials tried to push their communities to gradually reorganize certain common property, based on their communal temples, when the religious tradition gradually shifted from Vajrayana Buddha monasteries to village temples. By whatever means possible, the local elite, mainly the village gentries, searched for the common interests of the communities they belonged to. Moreover, the transportation route was the clue to link these elements together, such as the state power, local elite's influence, communal participation, and social integration, at the level of state politics as well as at the level of local cooperation.

As of the 1570s, the spatial circles around Zhaozhou city and the political hierarchy gradually built up, and this platform of connections between villages and county officials was also established to enable village gentries to share their village-based interests with other villages and county government, like the position of Flown Monastery in the northern basin. The monastery became a platform where village gentries met and planned

35 Yang and Hu, *Unofficial History*, p. 125.

36 Chen, *Jingtai Gazetteer*, p. 246.

their cooperation of local management and helped to deal with the state tasks in terms of transportation corvée.[37] Through the study of village temple inscriptions we found that villages were managed as a unit under their village leaders or gentries – official-title-holder scholars and retired officials – to work as representatives of their villages and to negotiate with the subprefecture government and neighbouring villages for irrigation channels and water sharing, social legal cases, and conflict. However, their activities were normally performed under the name of their village temples.

In general, there were three styles of temples in the Zhaozhou basin as of the 1570s. Firstly, the official temples controlled by county officials; secondly, the village temples controlled by village gentry; and thirdly, the public temples managed by a cooperative group of officials and gentries. The reconstruction movement of village temples coincided with the construction of the local space identified by the collective festival schedule of the whole basin, and it is another way for us to observe the social integration in basin society after the 1570s. All villages were organized by their village temples as the temples cooperated with each other as the connective sites of the irrigation system in rice farming in the basin, and this system was actually managed by village gentries under the hierarchy of a city-village-basin spatial arrangement. However, this spatial hierarchy was also applied through temple festivals in the basin based on an annual agricultural calendar. As a local custom, the streets in the county city organized their deity parade on the fifth day of the lunar New Year. Then a new yearly cycle started, followed by the parades of the village temple gods' birthday celebrations and large-scale temple fairs throughout the year. All these temple fairs and gods' birthday parades organized all communities, together, into a religious space as well as an agricultural circle, which showed their responsibilities and obligations to other, between village members and between village communities. The system of temple festivals therefore performed a crucial social-relationship role in the basin space.

Because of the transportation position on the main route that linked with Burma and with many prefectures in western Yunnan, after serious social chaos during the political transformation from the Ming to the Qing governments between the 1640s and the 1680s, and a flourishing economy during the Qianlong period (1735-1795), the transportation pressure had approached a new peak in western Yunnan. Many local labourers had to service the transportation of copper and silver and other supplies for mining industry development, as well as the long-distance trade between Yunnan,

37 Ma, 'Inscription on the Temple Property', p. 283.

Burma, and Tibet. The heaviest pressure on transportation came during the wars between the Qing state and Burma kingdom (1762-1770).[38] During this period, most villages established their common property to help deal with the government's tasks by taking their stable income from temple properties to pay off the state corvée tax, otherwise, an individual family could not afford such heavy pressures.[39] In this way, village temples gradually developed to become a kind of social institution for local communities to help them take responsibility for the state. Based on the working platform of common property around the village temples, local elite as gentries represented their communities to manage common income as a system of the proportion of corveé for individual families.

Reconstructing the Bazi geomancy and construction of local space

Since the 1380s, the officials had started to push the reconstruction of local geomancy order based on the theory and technique of *Fengshui* ideals, in order to reshape the local social landscape under the influence of Vajrayana Buddhism, which was controlled by noble and religious masters' families. From a series of stone inscriptions preserved in Flown Monastery we find that the ideal of making the Zhazhou basin into complete a *Fengshui* space started in the early Ming period. A piece of stone inscription says that, 'The Red Mountain is the water gate of the whole subprefecture, because it faces the Erhai Lake and the Cangshan Mountain, all historical literature is generated from this spirit of the lake and the mountain, thus this locality is the key and like the throat of a body.[40] And, because the Red Mountain was facing Flown Monastery, the main road and the market place were placed between them, and the lake piers were not far away, so the monastery was the exact gate of 'the Mountains of Phoenix'.[41] The ideal and image of 'the phoenix' constituted a fundamental part of the geomancy used by the Ming officials to reorganize the local landscape. The Ming government moved and rebuilt the county city to the western side of the basin, and this county had been named Zhaozhou for a long time.[42] This title means it was a territory subinfeuded to the Zhao family by the Nanzhao king, so the Zhao family was the master

38 Shilu, *Veritable Records*, p. 862.

39 Ma, 'Inscription on the Corvee Assignments', p. 250.

40 Ma, 'Inscription on the Temple Property', p. 283.

41 Ma, 'Inscription on Reconstruction', p. 285.

42 Ma, 'Inscription on the Construction', p. 230.

of this basin. However, the Ming officials tried to reinterpret the meaning of this space. The name of the mountain close to the newly built city was the Three Ears Mountain as it looked like three parallel ears, which represented the ear of heaven. After the city was moved to this location, a new name was given to the mountain: Phoenix Mountain (fengyi); the image of the three parallel ears was then regarded as the body of a phoenix in the middle, while two wings were equally positioned on each side. In order to match this *Fengshui* structure of a phoenix body, the county city was therefore gradually interpreted as the head of the phoenix, and the school in the city was the bird's mouth. The mountain facing Phoenix Mountain on the eastern side of the basin was regarded as Table Mountain. Finally, the school was designated as the school of the sound of the phoenix.[43] In addition, there was a gate of geomancy energy of this *Fengshui* space on the transportation route. This gate was also surrounded by two animal-like mountains: the mountain of the tortoise and the mountain of the snake, so the locality was also a platform of male-female (yin-yang) productive energy duality. The whole basin therefore became a site to promote the fortune of literati and civil examinations, and the forests on these mountains were well protected."[44]

Through the process of the continuing reconstruction of the Zhaozhou basin over several centuries, in the eyes of local residents and local officials, Zhaozhou was a complete space and a unique society because it had a perfect *Fengshui* structure, well-managed irrigation systems, a county city and government, an important transportation route bound with unknown barbarians, and the state capital at two ends. That made it a very crucial cosmology to the local residents in Zhaozhou basin. The names of mountains, rivers, temples, and festivals, also helped to reconfirm this cosmology of basin space and its surrounding mountains.

The road towards *All under Heaven* and the state power in the basins

The transportation system was run by labourers and their driven mules, horses, and vehicles. When this road became a core issue in state policies, there was no way to stop the transportation services on the route; it was only possible to maintain them through certain new methods. Under the promotion of local gentries, more and more people were willing to donate their property to support this project under the name of temple property,

43 Zhuang, *Wanli Gazetteer*, p. 19.

44 Ibid.

due to their religious belief. In this way, the title of 'merit rental' became a kind of religious merit for personal good even if it was used to deal with state tasks.[45] More and more villagers donated their property such as farming lands or mills to their village temples for religious 'merit making', and the rental income from the lands or mills was used to hire caravans for corvée services to perform the governmental tasks. In this way, the political responsibility thus transformed into a personal religious practice. The village temple, doubling as both a religious platform and a basic unit to perform the task assigned by the state, had effectively reinforced the development of local agency run by gentries, but one important driving force was the geographic feature of the basin on the state transportation route, and these basins were in fundamental positions to support the existence of the state power towards the southwest borderlands.

Returning to the early Ming period, when the Ming army occupied western Yunnan, the Ming government was not able to control this transportation route, and it took a very long time to guarantee the openness and safety of this transportation line from Burma to the capital Nanjing. The Ming emperors regarded this guarantee as the key issue of the 'face of the emperor' as the kings and chieftains had to be safe to pay their tributary to the emperor, and it meant that the frontier kings recognized the power of the Ming emperor.[46] Thus the cosmology of *All under Heaven* was accepted because the emperors were regarded as the 'son of Heaven' and had the ultimate authority in this world. Since then, the transportation route has been deeply embedded in the state power and bound with the reality of practical state cosmology.[47] In brief, the openness and safety of the transportation line were the foremost policies of the Ming and the Qing state. Along with the construction of *Fengshui* geomancy and the transportation framework, 'people could approach the Tibetans in the north, and approach barbarians in the west; to the south, the provincial capital could be approached, while to the east, the capital city of Beijing could be approached. For the exterior travellers, this is the road for their tributaries, while for the interior travellers, this is the road for official business and military generals.'[48] Due to the requirement of transportation, all logistics relied on labourers and mule caravans, but most of the government tasks were shifted to local communities to maintain the flows of materials and people on the road.

45 Ibid.; Chen and Li, *Daoguang Gazetteer*, p. 366.

46 *Related Documents*, pp. 30, 31, 438.

47 Ibid., p. 30.

48 Ma, 'Inscription of Hongshan', p. 325.

During the Ming and Qing dynasties, the most important policy towards frontier chieftains and kingdoms in Burma was to encourage them to pay an irregular tributary, through which the authority of Chinese emperors could be recognized, and the remote frontier people could be included into a framework of *All under Heaven* cosmological value. The tributaries of kingdoms in Burma, Siam, Laos, and Vietnam were therefore the most important state policies for officials in the southwest provinces. If these kingdoms seriously maintained their contributory relationship with the courts of the Ming and Qing dynasties, the borderland chieftains' territory would be peaceful. But sometimes wars between Chinese empires and these kingdoms occurred. Large amounts of military force were sent and war materiel followed if the central government could afford the cost of the wars. But in general, the main idea of this *All under Heaven* policy was to try to avoid war, by using the symbolic rituals to maintain the relationship between Chinese emperors and kings in Southeast Asia. Under this system, hundreds of local small or major chieftains had to pay their tributary all the time, and this political task alone made the transportation system flourish. For local officials and the communities along the transportation line, the task of protecting these tributary travellers and keeping them safe on their way, had been a most serious political duty and the highest state policy for local society, as it bound the state cosmology and political interest with identifying the power of the state and the travellers' recognition of the emperor as the son of Heaven.[49]

In summary, the construction of the Zhaozhou basin society represents a type of state existence in a mountain area, and it had taken dual responsibilities in the process of becoming a basin society: on one hand the villages in the basin had been well integrated in different layers as a local cosmological space, a cooperative agricultural system bound with the transportation system, and a local market shared with long distance trade; on the other hand, it had been a well-developed state institution since the 1380s as a social base to maintain the state transportation system, providing security and travelling services for guests of the state and travellers for different purposes.

Conclusion

When the political space was created and reshaped by the state, it also provided a chance for local agents, such as the leaders of local communities, to reshape their relationships with the state, and to reconstruct their

49 Ma, 'Inscription of Hongshan', p. 325; *Related Documents*, p. 438.

cultural meaning and life values. In this way, the idea of a road to a roadside community was similar to their understanding about the state.

Moreover, at the level of state politics, emperors and officials tried to build up and maintain the legitimacy of political power and its symbols through their official discourse about the relations between the Flourishing China (zhonghua) and the 'frontier barbarians (siyi)', to enhance the ultimate value of *All under Heaven* in everyday polices, through the symbolic journey of paying tributary, like a pilgrimage, from very far and remote places to the capital for a long time and undergoing arduous travel. This tributary journey was a powerful cultural symbol representing the serious sincerity and respect of remote chieftains to the power of the imperial states. In order to maintain this structure of state power, for the officials in Zhaozhou, their most important job was to support these travellers, receive them, and send them to another transportation recipient in another basin. Therefore, the idea and the symbol of *All under Heaven* was not an abstract caption, but a daily life practice for providing labourers, food, and vehicles. In ragged mountain terrain in Southwest China, especially in western Yunnan, transportation was a superior but also everyday task of politics, for the purpose of trade, official business, and tributaries paid by chieftains.

Even so, it is not an easy job at all, as endless rebellions and social chaos continually threatened all kinds of movement on the road. However, local communities had their own ways to respond. On one hand, an effective way of local agency could rapidly adjust its attitudes toward the state power, and adjust its cultural practice driven by intellectual elites like the gentries and religious masters; on the other hand, under the leadership of local gentries, different communities were still able to struggle for their collective space to reinterpret or recreate their cultural values. Different participants in basin society might have different purposes for these struggles, but what they learned was to make a collective village as a unique platform, to accumulate common willingness from the public in their competition with each other, and negotiation with the state. In this way, even if social relations may have varied, the connections became close as the village and the village leaders performed an important role of social agency for the basin society in the local, public sphere. In brief, the development of transportation networks supported by basin societies provides a new angle from which to understand the conditions of strong local agency for communal reconstruction for ethnic identities and political integration responding to state powers. The whole region of Southwest China and highland Southeast Asia could not be simply regarded as a homogenized terrain with the dichotomy of highland and lowland communities who could keep the state at a distance.

References

Chen, Wen. *The Jingtai Gazetteer of Yunnan (1455)*. Kunming: Yunnan Nationalities Publishing House, 2002.

Chen, Zhaotang, and Li Qixin. *The Daoguang Gazetteer of Zhaozhou Subprefecture (1839)*, vol. 3. Taipei: The Student Book Company, 1968.

Cheng, Jinren, ed. *The Qianlong Gazetteer of Zhaozhou Subprefecture (1736)*. Haikou: The Publishing House of Hainan, 2000.

Daniels, Christian, and Jianxiong Ma, eds. *Transformation of Yunnan in Ming China: From the Dali Kingdom to Imperial Province*. London: Routledge, 2020.

Dikötter, Frank. *The Discourse of Race in Modern China*. Hong Kong: Hong Kong University Press, 1992.

Elvin, Mark. *The Retreat of the Elephants: An Environmental History of China*. New Haven, CN and London: Yale University Press, 2004.

Fang, Guoyu. *Research on Historical Geography of Southwest China*. Beijing: Zhonghua Book Company, 1987.

Fuyi, Li. *The Letters of Fulu, Taipei*. Taipei: Furen Book Company, 2005.

Giersch, Charles Patterson. *Asian Borderlands: The Transformation of Qing China's Yunnan Frontier*. Cambridge, MA and London: Harvard University Press, 2006.

Giersch, Charles Patterson. 'Cotton, Copper and Caravans: Trade and the Transformation of Southwest China'. In *Chinese Circulations: Capital, Commodities, and Networks in Southeast Asia*, edited by E. Tagliacozzo and W-C. Chang, pp. 37–61. Durham, NC and London: Duke University Press, 2011.

Hechter, Michael. *Internal Colonialism: The Celtic Fringe in British National Development, 1536-1966*. International Library of Sociology. London: Routledge and Kegan Paul, 1975.

Herman, John E. *Amid the Clouds and Mist: China's Colonization of Guizhou, 1200-1700*. Cambridge, MA and London: Harvard University Asia Center, Harvard University Press. 2007.

Hostetler, Laura. *Qing Colonial Enterprise: Ethnography and Cartography in Early Modern China*. Chicago, IL and London: University of Chicago Press, 2001.

Hucker, Charles O. *A Dictionary of Official Titles in Imperial China*. Taipei: Reprinted by Southern Materials Center INC, 1985.

Leach, Edmund. *Political Systems of Highland Burma: A Study of Kachin Social Structure*. London: The Athlone Press, 1970 [1954].

Liu, Wenzheng, ed. *The Tianqi Gazetteer of Yunnan (1625)*. Kunming: Yunnan Education Publishing, 1991.

Lu, Jiarui, ed. *Gazetteer of Ethnic Minorities and Religions in Jianchuan County*. Kunming: Yunnan Nationalities Publishing House, 2003.

Ma, Cunzhao, ed. 'Inscription of Hongshan village'. In *Collection of Historical Inscriptions in Fengyi*. Hong Kong: South China Research Center, 2013.

Ma, Cunzhao, ed. 'Inscription on Reconstruction of Flown Monastery'. In *Collection of Historical Inscriptions in Fengyi*. Hong Kong: South China Research Center, 2013.

Ma, Cunzhao, ed. 'Inscription on the Construction of Zhaozhou City'. In *Collection of Historical Inscriptions in Fengyi*. Hong Kong: South China Research Center, 2013.

Ma, Cunzhao, ed. 'Inscription on the Corvee Assignments (1662)'. In *Collection of Historical Inscriptions in Fengyi*. Hong Kong: South China Research Center, 2013.

Ma, Cunzhao, ed. 'Inscription on the Temple Property of Wenchang Temple (1695)'. In *Collection of Historical Inscriptions in Fengyi*. Hong Kong: South China Research Center, 2013.

Ma, Jianxiong. 'Salt and Revenue in the Frontier Formation: State Mobilized Ethnic Politics in Yunnan-Burma Borderland since the 1720s'. *Modern Asian Studies* 48, no. 6 (2014): pp. 1637–1669.

Ma, Jianxiong. 'The Zhaozhou Bazi Society in Yunnan: Historical Process in the Bazi Basin Environmental System During the Ming Period (1368-1643)'. In *Environmental History in East Asia: Interdisciplinary Perspectives*, edited by Ts'ui-jung Liu, pp. 131–155. Abingdon and New York, NY: Routledge, 2014.

Ma, Jianxiong, and Cunzhao Ma. 'The Mule Caravans as Cross-Border Networks: Local Bands and Their Stretch on the Frontier between Yunnan and Burma'. In *Myanmar's Mountain and Maritime Borderscapes: Local Practices, Boundary-Making and Figured Worlds*, edited by Su-Ann Oh, pp. 237–257. Singapore: ISEAS Publishing, 2016.

Ma, Jianxiong. 'State and Reconstruction of the Southwest Frontier: Roads, Resources and Ritual Interpretations'. In *The Bazi Society of the Yun-Gui Plateau: Transportation Route, Resources and Ritual Interpretation*, edited by Z. Min and T-S. Liu, pp. 1–19. Kunming: Yunnan University Press, 2018.

Ma, Jianxiong. 'Local Communities, Village Temples and the Reconstruction of Ethnic Groups in Western Yunnan, 14th to 17th Centuries'. In *Transformation of Yunnan in Ming China: From the Dali Kingdom to Imperial Province*, edited by C. A. Daniels and J. Ma, pp. 43–75. London: Routledge, 2020.

Ma, Jianxiong. 'Constructing Native Chieftains as Imperial Frontier Institution: Endogamy and Dowry Land Exchange among the Shan-Dai Chieftains in Yunnan–Burma Borderland since the Thirteenth Century'. In *Yunnan–Burma–Bengal Corridor Geographies Protean Edging of Habitats and Empires*, edited by D. S. Yü and K. Dean, Ch. 8. New Delhi: Routledge, 2022.

Maxwell Hill, Ann. *Merchants and Migrants: Yunnanese Chinese in Southeast Asia*. New Haven, CN: Yale University Southeast Asia Studies, 1998.

Ni, Tui. *The Chronicle of Yunnan (1846)*. Kunming: Yunnan University Press, 1992.

Related Documents about Yunnan in Veritable Records of the Ming, Part 1. Kunming: People's Publishing House of Yunnan, 1959.

Scott, James C. *The Art of Not Being Governed: An Anarchist History of Upland Southeast Asia.* New Haven, CN and London: Yale University Press, 2009.
Shi, Shuo. *The Tibetan and Yi Corridor.* Chengdu: People's Publishing House of Sichuan, 2009.
Shilu, Gaozong. *Veritable Records of the Qing.* Gaozong Shilu, scroll 804, Vol.18. Beijing: Zhonghua Book Company, 1985.
Skinner, G. William. 'Marketing and Social Structure in Rural China: Part I'. *The Journal of Asian Studies* 24, no. 1 (Nov. 1964.): pp. 3–43.
Tong, Shaoyu and Yongsen Chen, *The Study of Bazi in Yunnan.* Kunming: Yunnan University Press, 2007.
Wang, Kangguo. 'Draft of Zhaozhou Gazetteer'. In *Collection of Historical Inscriptions in Fengyi*, edited by Ma Cunzhao. Hong Kong: South China Research Center, 2013.
Yang, Jiajing. *The Fengyi Gazetteer.* Kunming: Yunnan University Press, 1996.
Yang, Shen, and Wei Hu. *The Unoffficial History of Nanzhao* (1880). Taipei: Chengwen Publising House, 1968.
You, Zhong. *Collected Works of You Zhong*, vol. 2. Kunming: Yunnan University Press, 2009.
Zhao, Min. 'Salt, grain and the change of deities in early Ming western Yunnan'. In *The Transformation of Yunnan in Ming China: From the Dali Kingdom to Imperial Province*, edited by C. Daniels and J. Ma, pp. 19–42. London: Routledge, 2019.
Zhuang, Cheng, ed. *The Wanli Gazetteer of Zhaozhou Subprefecture* (1587). Dali: Reprinted by the Prefectural Bureau of Cultural Affairs, 1983.

About the author

Jianxiong Ma, Associate Professor in Anthropology, Hong Kong University of Science and Technology. His work targets the historical formation of the Sino-Myanmar frontier and ecological conditions of cultural diversity and ethnicity in Southwest China. His works include *The Lahu Minority in Southwest China* (2013) and *Reinventing Ancestors* (2013, in Chinese).

4 Tracking Routes

Imperial Competition in the Late-nineteenth Century Burma-China Borderlands

Gunnel Cederlöf

Abstract

For many past centuries, the Shan-Dai territories have been crossed by communication routes allowing low- and high-tide flows of pilgrims, diplomats, and the most valuable goods to move. This chapter in detail studies events in the late 19th century when British and Qing imperial forces sought to gain control of the movements and value of these flows. The effects of the Panthay Revolt in Yunnan together with a diminished Burmese kingdom resulted in unexpected opportunities for a British expedition to make their first journey ever between Bhamo in northern Burma and Momein (today: Tengchong) in Yunnan in 1868. By using the documentation generated by this expedition, this study traces a complex social web of relations that travellers on these routes had to negotiate.

Keywords: Burma, Yunnan, trade, borderlands, imperial history, flows

An unusual expedition, including an odd mix of men, travelled across the large and rugged mountain terrain between the Kingdom of Burma and the Yunnan Province of imperial China in 1868. It was the first British expedition to enter Yunnan and return to Burma by this route. Before setting off from Bhamo in north Burma, the political agent of Mandalay, Major Edward B. Sladen, who led the group, anticipated an easy walk that was made safe by Burmese royal letters. Sladen was soon disproved: the war between Burma and the British East India Company in 1853 had resulted in a large loss of land for the king of Burma, which he had refused to acknowledge by not signing the peace treaty. Bhamo was outside the British realms, and the British governor general discouraged Sladen, arguing that the enterprise

Cederlöf, Gunnel, and Willem van Schendel (eds), *Flows and Frictions in Trans-Himalayan Spaces: Histories of Networking and Border Crossing*. Taylor & Francis Group 2022
DOI: 10.5117/9789463724371_CH04

was too dangerous. However, by identifying the overland route into Yunnan, European merchants expected to get access to China's most lucrative markets, and the Rangoon Chamber of Commerce put in the necessary funds.[1]

In the mid- to late nineteenth century, three empires – the Qing, the British, and the French – were fast closing in on this mountain region. The Qing aimed to quell a rebellion in western Yunnan that had been under Panthay control since the outbreak in 1857, and the French wanted to beat the British in securing a route into Yunnan from their territories in Laos. Squeezed in between, the many small Shan-Dai polities in the mountain tracts came under immense pressure. Yet without their assistance, no one could make the journey across their territories. Sladen's fortunes depended on his ability to forge relations with the men controlling the route and its entry points, and this task was more complicated than he had foreseen. Before 1868, only a few brief journeys in the southern mountain tracts had been made in 1829-36.[2] No British reports on travel in the Shan-Dai territories reflected the dynamism of travelling, so Sladen lacked crucial information. Sladen and his party arrived at Bhamo with the royal letters in support of the expedition, but they were halfway into the hills before they realized that royal and imperial seals had little influence here. On their return from Yunnan, they believed they had secured future safe communications. They could not have been more mistaken; these territories were not the kind of channel for trade to which they only needed access. The social landscape was as complex as the physical one, and the grid of pathways fluctuated with political and environmental realities.

This chapter targets flows in order to enquire into how passage was made possible between the important hubs of wealth and markets in Burma and Yunnan. By studying the flows, we will better understand the social

1 I am immensely grateful to Jianxiong Ma for discussions and for sharing his work and experiences with me, including the translation of certain documents. I also wish to thank Christian Daniels and Dan Smyer Yü for their contributions to my knowledge of Yunnan and the Shan-Dai region, and to Dan for introducing me to Trans-Himalayan studies, sharing conceptual ideas, and joining me on fieldwork in 2018. I am also indebted to my colleagues in the 'India-China Corridor' project, Willem van Schendel, Mandy Sadan, Arupjyoti Saikia, and Dan Smyer Yü. My thanks also to Zhang Xilu, Li Donghong, Li Yunxia, Wang Yu, Zhengli Mu, and Nicole. The project has been funded by the Swedish Research Council. This study and its key arguments were presented in an early version of this text at the conference 'Flows and Friction in Trans-Himalayan Spaces', organized by the project at the Academy of Letters, History and Antiquities, Sweden, 2019. Comments from the participants have been helpful.

2 The officers Richardson and McLeod respectively travelled from Moumein at the mouth of the Salween River, Richardson to Laos, Siam, and Karenni in 1829, 1834, and 1836; McLeod to Laos, Siam, and Kengtung in 1836. Thaung, 'British Interest', pp. 48, 68, 119.

geography of mobility in the Shan-Dai region and its interdependence on and relationship with its powerful neighbours. The natural conditions were challenging for the passage of caravans, and the drivers required knowledge, experience, and skill. Yet they would not necessarily take the least physically demanding paths. Political realities, economic and kinship relations, and violent conflicts influenced which routes were chosen. As will be argued, while natural and social phenomena conditioned such a dynamic geography, flows and motion constituted it. The movement of people, animals, and items, and all the knowledge, relationships, status, and perceptions attached to them have nurtured the social geography of the Shan-Dai lands and merged with the social relations of places. In this perspective, flows are relative to space in contrast to being determined by absolute space. They intersect and integrate an anthropogenic and non-anthropogenic nexus.[3]

The extraordinary political circumstances during which the mission took place resulted in many reports and communications with dense narrative descriptions and strategic discussions. Here we can discern the complex negotiations and trivial disputes that determined the thoroughfare as seen from the perspective of the British members of the expedition and the merchant community and British civil and military administration with whom they communicated. British, Burmese, and Chinese officials also voiced their opinions about the state of affairs in various items of correspondence. The Shan-Dai polities, too, left traces in written documents, either indirectly when cited or referred to by the Qing, Burmese, or British administrations or, in closer encounters, in notes during this and following expeditions. Communication on route was filtered through translation, and the identity of the translators sometimes played key roles. There are also important collections of documents from the Shan-Dai polities' internal administration, which have survived in British archives. They were once taken as part of intelligence during British military expeditions to force the sawbwas to submit to British rule. For example, the James George Scott collection includes correspondence between the sawbwas in the southern polities who discussed between themselves strategies and alliances with which to confront the approaching British troops.[4]

To label the polities 'Shan-Dai' reflects their frontier character as represented in language. 'Shan' in Burmese and 'Dai' in Dai language denote the majority population. Also, before the Shan-Dai hills were divided by

3 This study and its key arguments build on Cederlöf, 'Tracking Routes'.

4 Scott was posted in the Shan polities (states) from 1886-1910, advancing from assistant commissioner to superintendent. Cambridge University Library, Manuscripts and University Archives, Scott Collection. The sawbwas' letters dated 1885-96 and many without a date.

a hard political border between Qing China and British-controlled Burma in 1900, these polities were under the influence of the Chinese Empire and the kingdom of Burma. Both governments appointed and recognized the heads of the polities – the *sawbwas* – within their realms, and they expected loyalty in return. An old border dating from the Ming Dynasty, marked by eight 'gates', had cut through the western part of the hills, but there was no hard political border. Seen across time, the reach of the two strong powers shifted. Sladen's records show how the shadow of the two giants faded away the further into the hills the caravan arrived.[5]

The Shan-Dai region had been embedded in state and imperial competition for centuries. Located in between the Irrawaddy River in Burma and Yunnan's high plateau, it hosted a web of communication routes. These were the millennium-old southwestern branch of the Silk Road network that connected Chinese trade with the Indian Ocean ports. The most valuable goods were transported in the north, between Bhamo on the Irrawaddy and Dali via Momein[6] in western Yunnan. Sladen called it 'a commercial highway'.[7] Research within the larger region, between northeast India and Yunnan via Burma, has generated important studies that have targeted the marginality and frontier character of tracts such as the Shan-Dai hills. Whether as a 'borderland', 'frontier zone', 'buffer zone', 'middle ground', or 'non-state space', their analytical location has depended on the faraway political centre of states. Even though these concepts share a critique of state-centred or nation-state-centred analyses by highlighting the peripheral areas as societies with their own centres and social life, the concepts identify them by their dependent relationship with a state, enclosing their geopolitical edge as a scar through the social body.

Jianxiong Ma and Charles Patterson Giersch engage in understanding this complexity in their research of the hill tracts. Both speak of 'borderlands' as regions without settled borders, which is a somewhat unusual definition within borderland studies in Asia that mostly presupposes a region cut apart by a hard political border. Giersch emphasizes the lack of political boundaries and argues that even though the Qing state claimed territories, there were no borders. For Ma, the formation of a frontier is the key question.

5 Coryton and Margary, 'Trade Routes', p. 267; Ma, 'Dowry'; Bowers, *Bhamo Expedition*, p. 7.

6 Momein in the Dai language was also Tengyueh in Chinese. Today, Tengchong. The British reports used the Dai name until the fall of the Panthay rebellion in 1873.

7 Oriental and India Office Collections, Selections from the Records of the Government of India, Foreign Department, LXXIX, Official Narrative of the Expedition to Explore the Trade Routes to China via Bhamo, by Major E. B. Sladen, 1868. Calcutta, 1870 (hereafter: OIOC, Expedition to China, E. B. Sladen, 1868), para. 26.

In contrast to Giersch, he claims that borders have been in place and have multiplied ever since the Qing state's bureaucratic reforms in the early eighteenth century. The Qing state's southwestern frontier was 'a crossroads and cultural transition ground between Tibet, [the] inner provinces in China, and Southeast Asian states'. With a long temporal perspective, Ma dates the beginning of the historical process that shaped the integration of the Shan-Dai polities into the Qing state to the 1720s and the conflicts over one of the most important minerals, salt. He argues that the mutually constitutive processes of state penetration and the local communities' reactions and their reconstruction of identities and cultural systems, together shaped 'frontier formation'. Ma makes a point of emphasizing this long-term integration and formation of cultural subjectivity in opposition to James C. Scott's claim that people escaped into the hills away from aggressive state encroachments.[8]

Giersch takes inspiration from North American scholarship when he introduces Richard White's term 'middle ground' into analysis of the Shan-Dai territories. The middle ground is a social rather than physical space, and Giersch focuses on the interaction within the frontier, among people and between them and the 'newcomers'. He writes that these are 'places of fluid cultural and economic exchange where acculturation and the creation of hybrid political institutions were contingent on local conditions.'[9] Richard White's term shares ideas with Mary Louise Pratt's concept 'contact zone' – their texts were published only one year apart. Pratt's concept is intended for a historicized approach to analysing travel writing, as 'social spaces where disparate cultures meet, clash, and grapple with each other, often in highly asymmetrical relations of domination and subordination'.[10] The 'contact zone' as the 'middle ground' aims to reveal the logic of meetings between two in all ways different people who need to find common ground in order to communicate, and therefore adjust and adapt. White's concept has a sense of manipulation, as when one party wants to subdue the other into meeting its own intentions, but lacks compelling force. Pratt's concept foregrounds 'the interactive, improvisational dimensions ... easily ignored or suppressed by diffusionist accounts of conquest and domination'. The study of such interactions is included in the meaning of Gunlög Fur's concept 'concurrences', when complementary or conflicting occurrences (as in meetings of counterparts) represent multiple voices, expressing 'concurrent

8 Ma, 'Zhaozhou Bazi Society', pp. 46, 131; Ma, 'Salt', pp. 1638-1639; Giersch, *Asian Borderlands*, p. 4; van Schendel, *Bengal Borderland*, pp. 1-4.

9 Giersch, *Asian Borderlands*, pp. 3-4, 7.

10 Pratt, *Imperial Eyes*, p. 4.

claims on geographical, temporal, political, and moral spaces.' They intersect at the same time and place, as rivals and competitors, representing different ontologies yet in communication with each other and they become entangled. All three concepts carry a dimension of meetings between people who are unknown to each other.[11] Ma, on the other hand, argues that the two – the people in the Qing state and the Shan-Dai polities – knew each other well from nearly two centuries of integration that involved adjustment and the formation of new identities on the frontier. Therefore, to him, 'middle ground' becomes a flat concept that cannot accommodate the historical dynamic of political resistance against the Qing state's encroachments in order to control local resources. If there ever was a middle ground of Qing and Shan-Dai encounters, it was long gone by the time of the first British attempts at entering the hills.[12]

A small observation should also be made on the perspectives represented in historical research. Until now, scholars have tended to enter into the study of the Shan-Dai region either from British Burma and its records or from imperial China and its subordinate governor's bureaucracies. The present nation-state border as well as language proficiency set limits. However, we have now reached a point where our individual limitations as scholars can be overcome by benefitting from the small but growing body of research from both sides of such boundaries.

Expanding imperial control

Qing and Burmese influence in the Shan-Dai region followed different trajectories. Burma's imperial ambitions, including wars on China in the north, Siam in the south, and Assam and Manipur in the west, came to a definite end when they were defeated by the British East India Company's (EIC) troops in 1826. In the nineteenth century, piece by piece, in three successive wars Burma lost its territories to the British. After the second war, in 1853, the EIC controlled all coastal areas and the delta almost as far north as the Burmese capital Mandalay. The company intended to secure land routes from India, across Burma, to form a stronghold in the markets in Yunnan. Qing rule in Yunnan had roots in the Ming administration, which formed tributary relations with the polities in westernmost Yunnan, west

11 White, *Middle Ground*, pp. 51-52; Pratt, *Imperial Eyes*, pp. 6-7; Fur, 'Concurrences', pp. 39, 40, 46.

12 Ma, 'Salt', p. 1640.

of the Ailao Mountain Range. For a long time, to the imperial rulers, these mountains marked a boundary between the state and the barbarian and wild tracts in the Shan-Dai territories towards Burma.[13]

The Ailao Mountains run northwest to southeast, from about 100 kilometres south of the commercial and political centre of Dali and 400 kilometres towards Laos. When the Qing state expanded beyond these mountains in the eighteenth century, its priorities were strategically guided by the location of mines, trade, and markets.[14] Through an administrative reform in 1746, counties were introduced into most of Yunnan where the region became known as the 'interior'. Exterior to this administration were the Shan-Dai polities where the native chiefs were replaced. These changes were paralleled by large-scale immigration of Han people into Yunnan that dramatically changed the demography, and the Han became the numerically dominant population in the province. In 1775-1850, Yunnan's population rose from four to ten million – an increase far above the China average. It worked to push native communities away from central Yunnan, westwards and southwards, beyond the Ailao Mountains and into the Shan-Dai region. The tensions that followed generated two large revolts, which resulted in the formation of polities that were autonomous of the Qing state. The Panthay revolt was crucial to Sladen's fortunes.[15]

The British intention to enter China from the western overland route was a consequence of the EIC's increasing control of territory east of the river Brahmaputra up to Burma's western borders – an area of small kingdoms and lively marketplaces. Back in 1817, Felix Carey in the service of the Raja of Cachar observed that high-value goods from Burma and China used to enter Bengal via Cachar. '… [C]ertainly then it must follow

13 As a criticism of dominant Chinese historiography that concludes that the war was a success for China, Yingcon Dai presents a radically different conclusion by comparing Chinese and Burmese correspondence and reports. Cederlöf, *Founding an Empire*, pp. 86-87, 91; Dai, 'Disguised Defeat', pp. 55-61, 66-67; Cederlöf, 'Seeking China's Back Door', pp. 130-133; Mangrai, *Shan States*, pp. 24, 44-45.

14 Ma, 'Rise'; Ma, 'Salt', p. 1648.

15 Taking account of the uncertainties of demografic statistics, James Lee estimates that the population in Yunnan increased from 4 to 10 million in 1775-1850. The annual growth of population density rose from an all-China average of 7/1,000 (1775) to 10/1,000 (1785) and 20/1,000 (1795). Industry and cities were the first to grow, and more periferal agrarian areas expanded later. Compared to decreasing mortality and increasing fertility, net immigration was largely the reason for the net increase in population. The two major revolts in Yunnan generated the autonomous regions of the Pignan State in the Panthay revolt 1857-73 and the Five Buddha districts in the Luohei Mountains, with a peak period in 1790-1890. Ma, 'Marriage'; Atwill, *Chinese Sultanate*, p. 6; Ma, 'Rise', pp. 5-6; Lee, 'Food Supply', pp. 729-731, 42; Ma, 'Shaping', pp. 70-71; Ma, 'Salt', p. 1639.

that these important articles of commerce, might be procured at a much cheaper rate than what we now get them from the Burmans who dispose of these articles to our merchants from their different sea ports, at a very enormous profit.'[16] When the EIC lost its two last monopolies in 1833, the *Charter Acts* that had given them sole rights to trade in tea and trade with China, access to this alternative route into China became urgent. Forcing the king of Burma to accept an agreement that protected and promoted trade between north Burma and the British Empire (1862) and, one year before Sladen's departure, an agreement with more far-reaching concessions that crippled Burma's control of trade, the commercial community in Rangoon acted even more aggressively. Through their initiatives to block American and French merchant interests, also without a peace treaty with Burma, the country was landlocked. Globally speaking, it was an unprecedented period of British imperial conquest.[17] The balance of power in the Burma-Yunnan borderlands did not shift dramatically until the end of the third Anglo-Burmese war in 1885. As a result, all Burmese territories came under British rule, and the powerful Qing and British empires came face to face in the intermediary Shan-Dai territories in the border negotiations that followed. Yet in spite of sovereign claims, neither the British nor the Qing state exercised control over the territories that separated them.[18]

The Shan-Dai region comprised a number of small polities that had their political and economic centres in flat oblong basins, *bazi*.[19] These basins were farmlands surrounded by mountains. A polity's main village, regular marts, and almost all towns were located here, whereas its realm extended east to west into the hills. Three main, parallel routes connected the lowlands at Bhamo with Yunnan's high plateau. The morphology made the basins run in the direction northeast-southwest, and the geological formation forced travellers to advance on the hillsides, along the watercourses in the narrow valleys, bordered by steep hill ranges on either side of the paths. All routes passed via valleys and basins, and passage, depended on the sawbwas. Without agreements confirmed beforehand, passage was hazardous. The

16 National Archives of India, FPP, 14 May 1832, No. 81, in Bhattacharjee, *Trade*, pp. 34-35.

17 Pollak, 'Origins', pp. 489-490, 502; Gallagher and Robinson, 'Imperialism', pp. 2-3; Myint-U, *Making of Modern Burma*, pp. 128-129, 36-37.

18 The officers Richardson and McLeod respectively travelled from Moumein at the mouth of the Salween River, Richardson to Laos, Siam, and Karenni in 1829, 1834, and 1836, McLeod to Laos, Siam, and Kengtung in 1836. Thaung, 'British Interest', pp. 48, 68, 119, 310.

19 See Ma's chapter in this volume, pp. 54-55.

routes crossed boundaries between different polities and alliances of which Sladen and his fellow officers were only vaguely aware.[20]

Ma's meticulous work on the organisation of the Shan-Dai polities reveals the unique integration of the exercise of imperial bureaucratic control with the marriage relations between the individual ruling families. Intermarriage linked the daughter of one sawbwa to the son of another, which reduced conflicts, especially because the sawbwa's wife controlled the imperial stamp. Without this stamp printed on documents, decisions were not valid. As noted, the polities were not ethnic entities. The British placed people within racial hierarchies, as closed ethnic communities according to the racial theories of the time, which caused misunderstandings.[21]

The Panthay revolt drastically changed conditions for travel, as the officers on route were slow to realize. Not only had the social organisation, tied into Qing administration, been cut off from the Qing state; with the Qing government replaced by the Panthay, Han Chinese merchants were unable to travel, and their merchant community in Bhamo was disconnected from Yunnan. Their connection with Yunnanese trade was now relocated further south, via Mandalay, and they sought support from the Burmese authorities. The Panthay revolt had a decisive influence on the success of Sladen's expedition to Momein – the headquarters of the Panthay governor, the *Tah-sa-kon* Lee Guo Lun. The expedition in fact took place at the peak of the rebellion. Close to its outbreak, in 1855, trade between Bhamo and Momein was calculated to be worth half a million pounds sterling – a substantial flow that had now receded to a mere trickle. The Burmese court feared that the British would enter Burma from its Indian territories in Assam and take control of Chinese trade via Bhamo. However, the revolt had put a complete stop to the caravans between Bhamo and Momein, and one of Sladen's tasks was to find out why.

The outbreak of the rebellion was preceded by a long and violent crescendo, caused by rising tension from the immigration of people from overcrowded villages in central China who established themselves in Yunnan's lucrative trade and mining business, and on agrarian lands. Whereas the Qing rulers were Manchu, most of the immigrants were Han. Han people had deep roots in Yunnan, too, but the forceful newcomers became an economic and political challenge to the native population, including the resident Han. Over time, localized Han lineage-associations formed a political system,

20 OIOC, Expedition to China, E. B. Sladen, 1868, paras 12-14. Ma, 'Zhaozhou Bazi Society'. OIOC, Expedition to China, E. B. Sladen, 1868, 30 May 1868.

21 Ma, 'Dowry'.

and a Han identity took shape. The Han merchants thereby found a way to control business across long distances. Before the rebellion, the Han merchant network connected places as far apart as central Yunnan and Mandalay via the intermediary Shan-Dai polities. A few decades later, the network had a global reach.[22]

The word Panthay originated in Burma, referring to Muslims living in Yunnan, where they were known as Hui. This community controlled the caravan trade on which merchants depended. The rebellion was ultimately triggered when violence between Han and Hui groups culminated in a massacre of the Hui population at Kunming in 1856. During the expedition's stay in Momein in 1868, the Panthay governor let Sladen know that Panthay forces had taken Kunming. Soon after the expedition left, the rebellion began to decline under the reinforced pressure of the Qing army.[23]

By the time Sladen set out from Bhamo, all trade flows had dried up. Blockages emerged from imperial competition, warfare, and uncertainties following the formation of the Panthay government in Dali. The Panthay now controlled the entry points in western Yunnan to the routes for the most valuable goods. In the longer perspective, lasting only fifteen years, the revolt can be seen as an exception in Yunnan's history. Paradoxically, it helps us to trace the missing flows and the mechanisms that controlled them. The hill societies were not immobile entities but were influenced by the flows of people, animals, and goods that crossed the hills and, on the way, integrated into markets and social life. Like seasonal clockwork, caravans in the long-distance trade and diplomatic missions passed through the polities. We need to observe how these flows of people and caravans – arriving, staying for a while, and leaving – impacted on the place. We can observe this, for example, in the case of one of the passages along the Bhamo-Momein routes that had the reputation of having places where 'trade flourished to such an extent that the caravan fires at the successive halting stations were never extinguished.'[24] We can sense the intensive life around those fireplaces, with the mule drivers and people from nearby places gathering together, eating, smoking, exchanging news, trading local items, haggling over prices, getting ready to sleep, or packing up to leave. The flows and the places mutually nurtured each other's social worlds.

22 Atwill, *Chinese Sultanate*, pp. 11-12, 145; Giersch, *Asian Borderlands*; Ma, 'Rise', pp. 3, 6.

23 Kunming, also known as Yunnan-fu. Anderson, *Report*, p. 328; Atwill, *Chinese Sultanate*, pp. 72-75, 90-99; Giersch, *Asian Borderlands*, p. 219.

24 Captain Sladen's Report, in OIOC, Expedition to China, E. B. Sladen, 1868, paras 153, pp. 191, 198, 200, 291, and 319, 28 May 1868, and 6 June 1868. Anderson, *Mandalay*, p. 38.

The caravan-flows moving from place to place and stopping at the caravan fires – these 'halting stations' also being the life worlds of people dwelling there, receiving caravans and seeing them off. To say 'halting station' is to keep to the social lexicon of the caravans. These were places for people on the move. They were also places for food vendors, craftsmen, local traders, and stable keepers. Products from the nearby villages and intra-regional trade arrived here too. The busy halting stations were constituent parts of towns or villages within the Shan-Dai polities and integrated into their socio-political organisation, the trade internal to the hills, the webs of kinship, and the religious universe.

Routes and caravans

Before we follow Sladen's trail, we need to identify the main routes between Bhamo and Momein. The northern path between the two towns was known as the Ponlyne or the Zanda[25] Basin route. It followed the Taping River, a tributary to the Irrawaddy that discharged into this river north of Bhamo. Following the Taping upstream, midway into the hills, the narrow valley in the lower hills opened up into the large Zanda Basin from where there was a steep climb towards the next basin and onwards to Momein. The Zanda route was separated by the Shamaloung mountain range in the south from the central Hotha Basin, known as the Embassy route. This route followed the Namsa River, a tributary that discharged into the Taping some 60 kilometres northeast of Bhamo as the crow flies, or about 80 kilometres on foot. The diplomatic missions between Yunnan and northern Burma travelled along this route. The third and southernmost route was the Sawadee route, named after the place on the Irrawaddy south of Bhamo, from where it originated. It followed the Nam-wan River upstream through the Muangwan Basin. Caravan drivers who chose this route would aim for Yongchang in Yunnan, and thereby never pass Momein on their way to Dali.[26]

The journey along the basins was physically less challenging than travelling in the intermediary forested valleys where the climbs were steep and paths narrow. Travel eastwards included a total climb of 1,500 metres from

25 (盏达), 'Sanda' in the British reports, today: Yingjiang.

26 OIOC, Government of India, Foreign Department, Political, No 39-41, 3 August 1868, Bhamo Expedition, Survey Report by Captain J. M. Williams, A.I.C.E., Engineer to the Expedition, 18 June 1868, para 8. OIOC, Expedition to China, E. B. Sladen, 1868, paras 16-18. Anderson, *Report*, p. 321.

Map 4.1 Communication routes between Bhamo and Momein, and the four zones, late 19th century

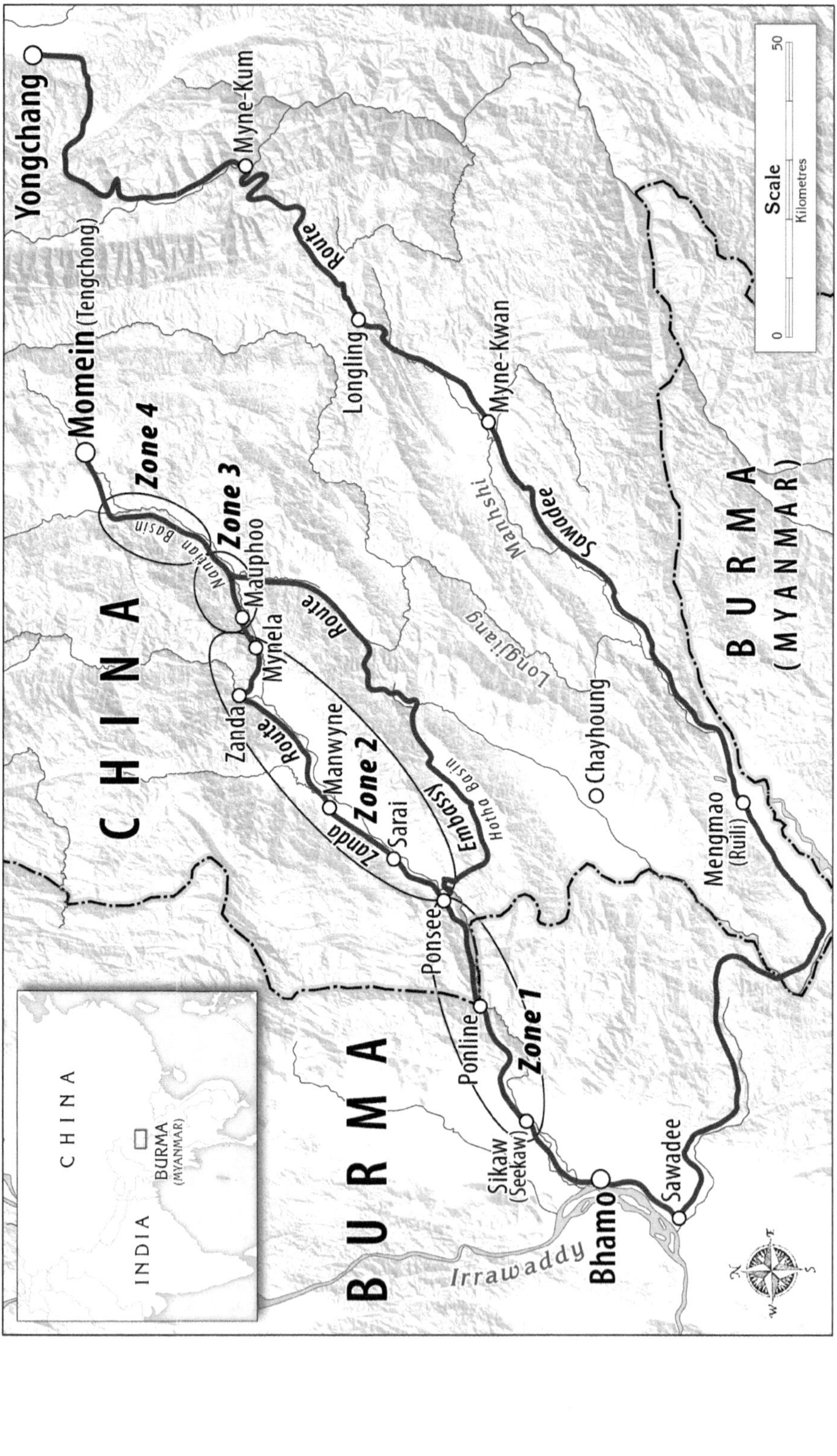

Drawn by Laurie Whiddon

Bhamo at about 100 metres above sea level to Momein at more than 1,600 metres. Comparing the British expeditions' notes on distances with satellite imagery may help explain the constantly higher numbers in the notes. Measuring stretches on foot across steep climbs may have exaggerated the geographical distances.

Nature contributed to facilitating and obstructing flows. These flows followed the seasons, travelling from October to March. In the summers, from April onwards, the routes were almost impassable, and the caravan drivers feared the fevers in the lowlands. The members of Sladen's expedition regretted many times that they had left Bhamo on 26 February to return on 5 September, travelling throughout the entire duration of the monsoon.[27]

The reason for being famed as the 'commercial highway' is reflected in estimates of the scale of the caravans. With the exception of the Panthay rebellion, 30,000 mules were reported to pass via this route every year. The caravans were led and sometimes also owned by a caravan headman, a '*ma-kuo-t'ou*' who paid muleteers, the '*mafu*', to drive the animals. Long-distance trade across Yunnan and into Laos, Siam, Burma, Sichuan, and Tibet depended on them. A caravan could comprise from 50 up to 500 mules. The more, the safer, for fear of robbery. One caravan assembled many small contracts and involved several mule owners. A few large owners in Yunnan possessed several hundred mules, whereas a common owner had between 70 and 80 animals. But this, too, was exceptional. Most owners were able to send their goods with 2-3 mules within a large caravan to a distant market. These caravans merged with the many travellers who moved goods within the mountains. They created a pulse through the basins, simultaneously being part of a larger, even global, world of commerce and competition, and being integrated into the social and economic life of the Shan-Dai polities. The communication routes as such were thereby deeply embedded in the society that channelled them. As in Martin Saxer's term 'pathways':

> A pathway is thus not just another word for trade route ... Life along a pathway is shaped by things, stories, rumors, and people passing through – by motion, or by flows, if you will. However, a pathway is neither just another word for flow. While shaped by motion, pathways are also conditioned by terrain, infrastructure and environmental factors like climate and weather.[28]

27 Willoughby, *Report*, p. 24; Anderson, *Report*, pp. 237, 386-87.

28 Saxer, 'Pathways', p. 105.

We need to observe that whereas the caravan drivers were mostly Hui and sometimes Han, the businessmen were often Han whose influence rested on their lineage associations. During the Panthay rebellion, these businessmen, such as the Yang lineage in Yun, were destroyed. The fact that they recovered quickly after the rebellion had been crushed reflected the strength of the political system represented by the associations. In the longer perspective, British rule in Burma contributed to transform western Yunnan and the Dai polities from being tributary to Qing rule to becoming dependent on the increasingly successful Han gentry and the British commercial entrepreneurs.[29]

The caravan economy also rested on breeding and keeping strong mules. They were known to be hardy, intelligent, docile, and well trained for pack work. Ponies were also bred in Yunnan, but they were not strong enough to walk in the rugged terrain in the western part of the province, loaded with heavy burdens. Nor would ponies last long in the heat of Burma's lowlands and they were therefore mainly used for riding. Mules on the other hand could carry up to 90 kilogrammes and walk stages of 16-40 kilometres before being replaced by fresh animals. Sladen and his companions experienced the muleteers' daily routine of feeding their mules before dawn with unhusked rice and thereafter watering them. After loading the mules, the caravan set off. At noon, the mules were unloaded, unsaddled, and turned loose to graze on the mountain side for an hour or so before entering the next stage of the journey. Once they had arrived at the day's destination, the animals were fed and watered, and either tethered close together for the night or stabled if a 'horse-inn' was available. Mostly, however, routines did not proceed this smoothly; delaying, haggling, and endless argumentation prevented departure.[30]

Four zones

Close observations of daily events during the expedition are available in the British members' diaries, reports, and correspondence. Sladen's report also includes a large number of incoming letters, including some from Burmese, Chinese, and Panthay officials. The British members were

29 OIOC, Expedition to China, E. B. Sladen, 1868, para 19. Willoughby, *Report*, pp. 20, 22-23; Ma, 'Rise', p. 11.

30 Examples of such delays are 1 and 2 March 1868, at Tsitkaw on the Taping River. Anderson, *Report*, pp. 242-243; Willoughby, *Report*, pp. 3-6, 22.

John Anderson, a medical doctor, naturalist, and the conservator of the Indian Museum at Calcutta; the engineer Captain J. M. Williams; the commercial representative captain Alexander Bowers; and the political agent of Mandalay Major Edward B. Sladen who was also in charge of the expedition. We may note the importance of these men, the empire's handymen, the soldier-merchants, the explorers, and the officers on scientific missions. Irrespective of whether their observations were right or wrong, they had bearing on the British government's judgements and decisions in India and Burma. The first expedition's reports became important guides for the subsequent one in 1875. Sladen's party consisted of 80-90 men including Theodore Stewart, a commercial representative who joined during the return journey; the interpreters F. N. Burn, Choung Zan Moung Mo, and Moung Shwe Yah; Jamadar Mahomed Ali Khan; soldiers of the Rangoon and Prome Police force; and the Shan caravan drivers and their mules.[31]

During the journey, Sladen's party encountered boundaries that we can think of as four zones. They were invisible to the survey members who struggled to understand the many obstacles on the way, but they were clear to everyone else. These were not fixed zones. As we shall see, they were temporally and spatially directed by social relations and natural conditions. Travelling from Bhamo, the party entered the first zone after a short boat ride up the Taping. The Shan-Dai polities were here under Jinghpaw sawbwas. Burmese influence in 1868 seems to have reached all the way up to the bottom of the Zanda Basin. This was where they entered the second zone. From here on, the sawbwas were Shan-Dai, and the Qing state had historically strong relations to the polities. The third zone began with a steep ascent up to Mauphoo (also: Maofu) and reached as far as the bottom of the next basin, Nantian. The war had created this zone, and the polities were headed by members of the Han community. Originally, they were Ming garrison-settlements that had entered the organisation of the Shan-Dai polities under Qing rule. They were now loyal to the Qing state and they sided with the Qing and the Burmese rulers against the Panthay and their allies.[32] The fourth zone comprised the Nantian Basin up to Momein. This zone was controlled by the Panthay. We will return to these zones in a while.

31 Coryton and Margary, 'Trade Routes', p. 267; Ma, 'Dowry'; Bowers, *Bhamo Expedition*, p. 7.
32 云南分县详图1941年8月版, Yunnan Sub-county Detail August 1941 edition. Captain Sladen's Report, in OIOC, Expedition to China, E. B. Sladen, 1868, para 66, 79, and 14 March 1868 in conversation with the Ponlyne sawbwa. Ma, 'Dowry'.

One of the grave mistakes the British officials made was to assume the existence of long-term stability and fixed relations between the Burmese and Qing states, and the Shan-Dai polities. Occasionally, the notion of the 'princely states' in India resonates in reports where the polities are called 'states'.[33] They were certain of the Panthay government's longevity and calculated how to accommodate trade with the Panthay regime in the long term. British-authored documents do not reflect any significant awareness of the micro-politics with shifting loyalties and changing relations of power that were influenced by disputes in the hills and the lowlands. In Burma's eastern lowlands, conflicts between Burmese officials and Shan and Jinghpaw sawbwas often escalated into violence. The lowlands were frequently raided, and Jinghpaw people were not welcome in Bhamo. They were disarmed at the city gates, and if they had come without permission, they risked getting killed. The Jinghpaw were also prevented from performing their religious ceremonies in and near Bhamo. Sladen noted that selling and buying bamboo, bullocks, and buffaloes that were necessary for an important offering, were barred. Apparently, the Jinghpaw experienced boundaries that targeted them selectively, and they responded by not following orders that originated from among the Burmese.[34]

The British, too, were not welcome in Bhamo, but in contrast to the Jinghpaw, they were treated with a feast, invited by the Chinese merchant community. Still ignorant of the hurdles piling up before them, they arrived into Bhamo harbour in grand style but found the town a ruin of its former prosperity. The Chinese community was disconnected from its closest link to Yunnan, and the absence of long-distance trade had reduced the town to poverty. Neither the Burmese government officials nor the Chinese merchants wanted British competition in the Yunnan trade, so they discouraged them from entering the hills. It was said to be a 'wild goose chase' with hill tracts filled with armed robbers.[35] At the party, Sladen was introduced to one of the Jinghpaw sawbwas. The Ponlyne sawbwa had remained silent as he was forced to act as a Burmese pawn. But once on their own, the

33 Until the British conquered all of Burma, they argued that the Shan states were sovereign as a buffer against Qing China. This terminology disappeared as soon as the polities came under British rule.

34 Kachin spelt 'Kakhyen' in the notes. Captain Sladen's Report, in OIOC, Expedition to China, E. B. Sladen, 1868, paras 39-42, 58, 62-63.

35 Captain Sladen's Report, in OIOC, Expedition to China, E. B. Sladen, 1868, paras 36-38, 45-48, 77-79, 89, Appendix, Letter from Captain E. B. Sladen, Political Agent, Mandalay, to Colonel Albert Fytche, Chief Commissioner, British Burma, and Agent to His Excellency the Viceroy and Governor-General, Bhamo 22 January 1868, p. ii.

sawbwa promised Sladen that he would procure mules and find caravan drivers locally, in the Shan villages.[36]

As they entered what we call the first zone, Sladen realized that here, away from Bhamo, the Jinghpaw sawbwas made decisions independently of the Burmese, and the deference he had witnessed disappeared. In these lower hills, the sawbwas were Jinghpaw who had adopted the hierarchical social organisation of the Shan-Dai polities. Mandy Sadan explains how the Jinghpaw areas in north Burma that were dominated by the Lahpai lineage, as part of a larger movement of people eastwards, extended their territories into the border zone of Yunnan to become entangled in the Panthay revolt. Sladen travelled in the company of the three most influential sawbwas, those of Ponlyne, Ponsee, and Saray. Day after day, Sladen complained about how the Jinghpaw cheated them, raised the price, and disappeared with the mules to enforce their unfair claims. His indignation seems rooted in a contractual logic, where the sawbwas broke agreements. However, in the first zone, the sawbwas acted as autonomous heads of polities both in relation to foreigners and to people subordinate to them. In the diary, disappearing mules, inflated mule hire, and vanishing luggage were recurring themes.[37]

Once they reached Ponsee, not far from the Zanda Basin, the expedition came to a complete halt. The sawbwas positively refused to bring them one step further. Burmese influence faded away at this point, and Jinghpaw sawbwas could not control the party inside the Zanda Basin. Sladen's party was stranded for two months without guides, and all the mules were gone. Unable to understand why they could not advance, he wrote in his diary: 'The proofs which I was able to produce in the King of Burma's written proclamation, and the letters given to me by the Burmese Government previous to my departure from Mandalay, were as waste paper compared with facts as disclosed by Burmese officials themselves in our immediate neighbourhood.' Away from Bhamo, Burmese representatives were loyal to the interests of the Governor of Bhamo, not to the king. Additionally, the Jinghpaw sawbwas warned Sladen about the 'most infamous Chinese robber, Li-Hsieh-Tāi', waiting for him just beyond Zanda Valley. The governor had ordered them to delay the mission and shirk their promises and, if

36 OIOC, Government of India, Foreign Department, Political, No. 39-41, 3 August 1868, Bhamo Expedition, Survey Report by Captain J. M. Williams, A.I.C.E., Engineer to the Expedition, 18 June 1868, para 10, 12. Captain Sladen's Report, in OIOC, Expedition to China, E. B. Sladen, 1868, paras 49-51, 104, 106; and letter from Captain E. B. Sladen, Political Agent, Mandalay, to Colonel Albert Fytche, Chief Commissioner, British Burma, and Agent to His Excellency the Viceroy and Governor-General, No. 194, Mandalay, 25 September 1868, para 5.

37 Sadan, *Being and Becoming*, pp. 55, 103-106, 78-79.

necessary, kill the mission's members. Yet even though Sladen elaborated on the possible alliances between the Jinghpaw, Burmese, and Chinese parties, he associated most of the daily difficulties with a lower level of development among the Jinghpaw compared to the Shan as between two separate ethnic groups. Not only were the Jinghpaw claimed to be treacherous when breaking agreements, they were also associated with dirt, drunkenness, and immorality. He never compared the relative poverty of the villages in the hills to those in the basins where there were cultivated fields, marketplaces, and more affluent temples and buildings.[38]

The influence of the translators' work is evident in Sladen's and Anderson's reports. The word used to denote the Jinghpaw community was 'Kakhyen', in today's transcription: Kachin. The British officers adopted not only this word from the non-Jinghpaw translators but also their explanations. The word 'Kakhyen' indicated barbarian people in forests and mountains, a people without civilisation. The one exception is when Anderson explained the language of some Jinghpaw men in the Zanda Basin, calling them Khung (Khaang), the Dai word for Kakhyen. The interpreters became filters in conversations and were in no way neutral to the expedition. Choung Zan Moung Mo, for example, was a village headman who had been assigned by the Burmese governor of Bhamo.[39]

The long halt at Ponsee is largely explained by the location of the place. The geography had political significance. Here the route from Bhamo arrived at a bifurcation with the Zanda Basin only a few kilometres ahead and the Embassy route turning towards the southern point of the Hotha Basin. The Panthay government had a hold on this route via the Hotha sawbwa, who had close relations with the Governor of Momein Lee Guo Lun – the sawbwa himself having ancestors in Yunnan. Ponsee's importance did not only stem from its location at this intersection; this was also the place where high-value merchandise balanced between Burmese and Panthay control. As soon as the Panthay governor of Momein realized that the Burmese were close to succeeding in blocking the British from reaching Momein, he ordered the sawbwas to aid the expedition and to take the Embassy route.

On seeing the turn of events, the Jinghpaw sawbwa of Saray grasped the opportunity and struck a personal deal with Sladen for escort, porters, and

38 Captain Sladen's Report, in OIOC, Expedition to China, E. B. Sladen, 1868, paras 36-38, 45-48, 77-79, 89, 129, Appendix, letter from Captain E. B. Sladen, Political Agent, Mandalay, to Colonel Albert Fytche, Chief Commissioner, British Burma, and Agent to His Excellency the Viceroy and Governor-General, Bhamo 22 January 1868, p. ii.

39 Ibid., para 165. Daniels, 'Blocking', p. 135.

mules. He also kept a line of retreat open by sending his Chinese interpreter to Bhamo to find out the consequences of assisting the British party, in opposition to both Burmese and Chinese orders. The response was resolute. In the early morning, one week later, two Chinese men from Bhamo passed through the village together with the Bhamo government's secretary, the sawbwa's Chinese interpreter, and a party of fifty armed Burmese soldiers aiming for Saray. They claimed to be on their way to inspect silver-mining operations. It was, however, a powerful demonstration of the Burmese governor's and the Chinese merchants' hostility towards the expedition. On the following morning, the Saray sawbwa came to explain that he was prevented from giving any assistance. The Burmese officers sent small groups of soldiers into the Shan villages to condemn the mission and paid bribes to Jinghpaw men to carry off the ponies. The wife of Sladen's interpreter was held hostage at Bhamo to force her husband to leave the expedition. Meanwhile, the letter from the governor of Momein disappeared together with Sladen's clerk. The absconding clerk was later promoted to the position of Burmese royal detective. All sawbwas now turned against the mission, and Sladen saw the glue that held his expedition together beginning to dissolve. Soon he realized that there had never been much of a glue in the first place. The Saray sawbwa explained: 'You will get no assistance from Burmans: they have repeatedly sent messages to tell me that you are not to go on. If you persist, we have been advised to make away with you in any way we please, and the reward offered is the appropriation of your property.'[40]

In this turmoil, Sladen kept to his wait-and-see strategy, and, eventually, it paid off. The Panthay governor sent a message that he intended to meet Sladen in person at the northern end of the Zanda Basin. This sufficed to convince the sawbwas to allow the expedition to advance, as they knew the trying consequences of having Panthay troops passing through their villages. They blamed the past fifteen years' fights and raids on the Panthay and were eager to have the trade route re-opened. The Panthay governor, too, wanted trade re-established and a trade agreement with the British drawn up. And so, Sladen was able enter into the second zone, the Zanda Basin.[41]

40 The government secretary at Bhamo, the Tsayay-daugyee. Captain Sladen's Report, in OIOC, Expedition to China, E. B. Sladen, 1868, paras 103, 119-120, 128-133, 137, 9 March, p. 56, and Appendix S. Translation of a statement made by Moung Moh, Kakhyen Interpreter, resident of Bhamo, on 27 November 1868.

41 The stay at Ponsee lasted from 7 March to 11 May 1868. Anderson, *Report*, pp. 87, 252-253.

Competition for the commercial flows

Zanda had a long history of imperial administration, dating back to the fifteenth century. Arriving from the southwest, a traveller passed through a well-ordered society, and, unlike the northern part of the basin, it was untouched by the war.[42] '[T]he whole valley area teemed with villages, and was alive with a population which had laid out and conjoined every available acre into one vast garden of fertility and wealth', according to Anderson. People were mainly Shan, Han, and Jinghpaw, and the mix of languages spoken in the basin reflected the intense commerce and flow of goods and people that connected the basin with neighbouring and distant places. Flows of refugees had arrived from places of war in Yunnan. Describing 'a perfect babel', Anderson listed Shan, Poloung, Lisu, a form of Mandarin Chinese spoken in Yunnan, and a Jinghpaw dialect with 'a remarkable affinity to Burmese' spoken by the Khung (Jinghpaw). The continuous movement of people, animals, and goods not only nurtured commerce but constituted flows of knowledge, ideas, and associations of all kinds. Political and religious ideas and affiliations travelled both with diplomatic missions and with the nuns in the powerful sawbwas' wives' courts. Sladen always asked for an audience with the *sawbwa gadaw* in each polity's main town. These flows made a powerful mobile social geography.[43]

Sladen calculated the size of the population from the number of households. Larger towns amounted to 1,200 houses and average villages 50. The sawbwa of Mynela in the south would thereby be the head of about 25,000 individuals, and the ten polities in the northern Shan-Dai territories would together amount to a quarter of a million people. The figure may give us an indication but is quite unreliable. Imperial influence was present in the payment of tribute. Mynela annually sent 5,000 baskets of paddy to the Panthay government – an amount equal to what they paid under the Qing regime.[44]

42 Ma, 'Dowry'. Captain Sladen's Report, in OIOC, Expedition to China, E. B. Sladen, 1868, para 208.

43 Muangla, 18 July 1868, Anderson, *Report*, p. 347. Captain Sladen's Report, in OIOC, Expedition to China, E. B. Sladen, 1868, para 153, 165, 228, 297. Coryton and Margary, 'Trade Routes', p. 289; 'Medical and Physical', p. 25.

44 The Northern Shan polities amounted to ten according to the Shan and eight according to the Burmese. The eastern ones were Mynechai, Mynechon, Sayfan, Mynemaw, and Mynewun, and the western ones were Hotha, Latha, Mynela, Zanda, and Mynetee. They were located in four basins: (1. the Zanda Basin) Mynela, Zanda, and Mynetee, (2. the Hotha Basin) Hotha and Latha, (3. the Mynewan Basin) Sayfan and Mynewun, and (4) Mynechai, Mynechon, and Mynemaw. A sawbwa was described as having a government of four men who drew a fixed

Sladen's message to the sawbwas was that the British monarch, represented by a minister at the court of Peking, would address the imperial government for the restoration of trade. Panthay sovereignty ought to be recognized by the Qing, which would secure peace and restore Shan prosperity. He further claimed that it was 'not uncommon' for Shan sawbwas in states adjacent to British Burma to submit to British rule and, immediately contradicting himself, that the British were not interested in more territory but in trade. His ultimate promise was to influence the Panthay governor not to attack the Shan polities anymore. An overly optimistic and not entirely true assessment of the situation.[45]

The polities in the different basins competed for the trade flows, and the Zanda Valley's neighbour, the sawbwa of Hotha, was eager to establish the Embassy route as the only highway. Once Sladen was set to leave Zanda, the Hotha sawbwa grasped the opportunity and turned up with 150 mules loaded with cotton for Momein. He joined the caravan and kept Sladen busy with descriptions of the Hotha Basin's commercial advantages and shorter route, and the dangers of choosing the Zanda or Sawadee routes. On Sladen's return journey through the Hotha Basin, the sawbwa even suggested a system of credits for the duration of the war. Sladen concluded that that route would make the Hotha sawbwa chief trader and he suspected double-dealing when realizing the sawbwa's involvement in the conspiracy to destroy his expedition at Ponsee.[46]

Without knowing it, the Europeans got a first glimpse of the intra-hill control of trade when they reached Karahokha,[47] a town at the northern end of the Zanda Valley that was under the control of the sawbwa of Nantian, as a branch of that polity. They noticed the large Chinese population and the lively central mart where the road was flanked by Chinese shops. Fully up and running, the whole roadway would be occupied by stalls in which butchers, bakers, druggists, and jacks-of-all-trades ran thriving businesses. But their Shan guides steered them away from the town, not mentioning its link to Han commerce, only claiming the caravan would get stuck in the crowds.[48]

form of revenue from certain villages, each with a headman (tamon). Captain Sladen's Report, in OIOC, Expedition to China, E. B. Sladen, 1868, paras 104, 208, 216, 219, and 270.

45 Ibid., paras 238-239, 389. Coryton and Margary, 'Trade Routes', p. 267.

46 Captain Sladen's Report, in OIOC, Expedition to China, E.B. Sladen 1868, paras 372-373, 402, and App. P. Report of Assistant Surveyor Moung Thazanoo, on the southern or Sawuddy route across the Kakhyen hills between Bhamo and Mowun, 31 August 1868.

47 Also Taiping Jie (太平街). Information about Karahokha has been provided by Jianxiong Ma.

48 Captain Sladen's Report, in OIOC, Expedition to China, E. B. Sladen, 1868, paras 229-230.

Ever since Bhamo, reports had arrived about the movements of Li-Hsieh-Tāi, his name meaning Brigadier Li. He was one of the seven chieftains, or Han sawbwas, close to Momein who were loyal to the Qing state and had rebelled against the Panthay regime. They controlled the hill passes from the Zanda Basin up to the higher placed basin, Nantian. Sladen soon adopted the terminology and called them alternatively 'robber chiefs' or 'Chinese brigands'. Mauphoo was one of the natural bottlenecks on the route, which could only be reached after a steep climb up a narrow valley. It was a customs point under Qing authority and, before the Panthay captured Momein, it was linked to this town. This was the place of such intense caravan trade that the fires were never extinguished. Geography and politics here also applied a squeeze on the trade flows through hefty duties or, as now, caused a complete blockage.[49] However, the Panthay governor made short work of the ambush party, eager to get Sladen to Momein. When the caravan finally arrived at Mauphoo, the bodies were still scattered on the ground from when the Panthay had attacked with 5,000 men, claiming to have killed 300 and driven away the others. Sladen's confusion was complete when they entered Nantian and were entertained as the governor's guests by two of the Han sawbwas who had quickly transferred their allegiance to the governor. Later, when the Panthay were defeated, these Han sawbwas returned their loyalty to the Qing government and Li-Hsieh-Tāi was appointed Commissioner over the Shan-Dai polities.[50]

After uneventful days crossing the Nantian Basin they arrived at Momein, somewhat reduced due to robbery but received in grand style. Governor Lee Guo Lun went out of his way to show support for the British plans to reopen the trade route and connect commerce in Momein with the British ports at Rangoon and Pegu. Sladen's assessment of the Panthay was positive and optimistic. Violence in Yunnan was blamed on the Qing, who were called marauders who harassed and molested the peaceful Panthay. On leaving, Lee Guo Lun and Sladen exchanged official seals. Sladen wrote that using them would 'give an impress of reality to our correspondence and secure us mutually against fraud and imposition'.[51] However, he should have paid

49 Ibid., paras 153, 191, 198, 200, 291, and 319, 28 May 1868 and 6 June 1868. Anderson, *Mandalay*, p. 38.

50 The other two were Leoq-wanfan, the most powerful of seven rebelling polities, and Thoung-wet-shein, the head of Mauton between Nantian and Momein. Captain Sladen's Report, in OIOC, Expedition to China, E. B. Sladen, 1868, paras 152-155, 290. Coryton and Margary, 'Trade Routes', pp. 289-290. Information on Qing customs points from communication with Jianxiong Ma.

51 Captain Sladen's Report, in OIOC, Expedition to China, E. B. Sladen, 1868, paras 319, 325-329, 6.6, 6.7, and 30 June 1868, and pages 115 and 160.

more attention to the governor's preparation for their departure. Messages went out to remove blockages on the road and to check for information of ambush parties. The sawbwas of Nantian, Mynela, Zanda, and Hotha received instructions to prepare for the expedition's arrival, and the sawbwa of Hotha again accompanied them and steered them into the Hotha Basin. Three hundred Panthay soldiers escorted the party as far as Mynela. Sladen may have succeeded in reaching Momein and returning to Bhamo, but he failed to break the code of how to do it. Members of the next expedition in 1875 paid with their lives for this omission.[52] At that time, the British officers saw the growing trade flows on the Bhamo-Momein route, after the suppression of the Panthay, as an opportunity. Chinese firms in Rangoon established branches at Bhamo, and the Burmese government ran steamers between this port and Mandalay. The recently sanctioned Rangoon and Irrawaddy State Railway was regarded as proof of the British government's forceful initiative to reach the inland markets, and the route was 'destined to develop far beyond its condition at any previous point in history.'[53]

But the British failed to understand the effects on the Shan-Dai communication routes of having the Qing government and, in effect, the Han merchants back in power in Yunnan. In 1875, the king of Burma first advised the new expedition to take the Bhamo-Momein route, but when he received news that the Qing government wanted renewed diplomatic ties, he told the British to cancel the journey altogether. When they nonetheless went ahead, arguing that they already knew the route, a harder border between Burma and China had arisen in the hills, and the mission became fatal.[54]

However, in 1875, this border was still negotiable, and it would continue to change. It was not until the geopolitical border between China and British Burma was fixed in 1900 that the Shan-Dai polities' room for negotiation crumbled. Analytically, the modern notion of hard nation-state borders that are characteristic of many important works in borderland studies is of no help here. However, we may ask whether the complex events that foreshadowed the hard border of 1900 may contribute a more dynamic approach in borderland analysis. The Shan-Dai polities were not a borderland, as an entity, cut apart by a geopolitical border, but a complex society of dynamically altering borders in the plural form, selectively creating obstacles to people travelling here. A traveller had to navigate the intertwined relationship of social and

52 Ibid., 6 June 1868 (p. 134); 7 July 1868 (pp. 143, 146). Coryton and Margary, 'Trade Routes', pp. 289-290.

53 Coryton and Margary, 'Trade Routes', p. 270.

54 Thaung, 'British Interest', pp. 484-490.

natural geographies – the anthropogenic and non-anthropogenic conditions. The jagged morphology forced flows in particular directions, and the dry and wet seasons conditioned the annual low and high tides of long-distance communication. Similarly, the socio-political life in the separate Shan-Dai societies influenced the flows since the sawbwas controlled the basins through which the flows passed. The basin societies were also interdependent through kinship and competition. Simultaneously, the polities had to balance the pressures of the stronger powers – Qing China, the kingdom of Burma increasingly replaced by British Burma, and, temporarily, the Panthay regime. These anthropogenic and non-anthropogenic circumstances continuously changed over time, which simultaneously altered the conditions for the region's social geography that channelled flows.

The British had a very selective reason for chasing information about the flows: they were seeking the most valuable current, only having eyes for profitable goods on a world market. However, they noticed in passing the many other currents, such as the trade internal to the hills, the religious institutional and theological influences as in the well-travelled and knowledgeable nuns, the inter-marriage of sawbwa families, and the devastating effects of armies. Material and immaterial value of the flows integrated into the basin societies, nurturing and altering them. Simultaneously, value and politics originating within the polities also joined the flows along the communication routes and spread across short or long distances. Information flowed quickly between the basins as the sawbwas' correspondence in the southern polities show, and the sawbwas tried to attract the valuable commodity flows into their own realms. The absence of flows during the Panthay-Qing conflict drained the northern Shan-Dai societies of energy. When we follow the flows, the analysis will not be determined by the social or natural geography or the place, but it will observe how flows contributed to constituting places.

The obstacles, blockages, and friction, causing flows to slow or pass in other directions, stand out in situations when the most valuable flows disappear from a particular area. Sladen's expedition entered right into such a void. They struggled daily to negotiate with their guides but occasionally hit a wall that their guides were also unable to pass. Such boundaries are explained by the alliances that crossed the social geography. Sawbwas negotiated alliances with each other and were under the influence of the neighbouring strong powers. Both the kingdom of Burma and the Qing state controlled the sawbwas by demanding services and tributes. When the Panthay entered into this equation, they tried to adopt the Qing administration's procedures. It resulted in dynamically changing zones of control and influence.

References

Anderson, John. *A Report on the Expedition to Western Yunan via Bhamo*. Calcutta: Office of the Superintendent of Government Printing, 1871.

Anderson, John. *Mandalay to Momien: A Narrative of the Two Expeditions to Western China of 1868 and 1875, under Colonel Edward B. Sladen and Colonel Horace Browne*. London: Macmillan, 1876.6

Atwill, David G. *The Chinese Sultanate: Islam, Ethnicity, and the Panthay Rebellion in Southwest China, 1856-1873*. Stanford, CA: Stanford University Press, 2005.

Bhattacharjee, Jayanta Bhusan. *Trade and Colony: The British Colonisation of North East India*. Shillong: North East India History Association, 2000.

Bowers, Alexander. *Bhamo Expedition: Report on the Practicability of Re-Opening the Trade Route, between Burma and Western China*. Rangoon: American Mission Press, 1869.

Cederlöf, Gunnel. *Founding an Empire on India's North Eastern Frontiers, 1790-1840: Climate, Commerce, Polity*. New Delhi: Oxford University Press, 2014.

Cederlöf, Gunnel. 'Seeking China's Back Door: On English Handkerchiefs and Global Local Markets in the Early Nineteenth Century'. In *Trans-Himalayan Borderlands: Livelihoods, Territorialities, Modernities*, edited by D. Smyer Yü and J. Michaud, pp. 125–144. Amsterdam: Amsterdam University Press, 2017.

Cederlöf, Gunnel. 'Tracking Routes: Imperial Competition in the Late 19th Century Burma–China Borderlands'. In *Flows and Frictions in Trans-Himalayan Spaces*. Conference held at the Royal Swedish Academy of Letters, History and Antiquities in Stockholm, Sweden, 2019, unpublished.

Coryton, J., and A. R. Margary. 'Trade Routes between British Burmah and Western China; with Extracts of Letters from Mr. Margary'. *Proceedings of the Royal Geographical Society of London* 19, no. 4 (1874–75): pp. 264–291.

Dai, Yincong. 'A Disguised Defeat: The Myanmar Campaign of the Qing Dynasty'. *Modern Asian Studies* 38, no. 1 (2004): pp. 145–189.

Daniels, Christian. 'Blocking the Path of Feral Pigs with Rotten Bamboo: The Role of Upland Peoples in the Crisis of a Tay Polity in Southwest Yunnan, 1792 to 1836'. *Modern Asian Studies* 2, no. 1 (2013): pp. 133–170.

Fur, Gunlög. 'Concurrences as a Methodology for Discerning Concurrent Histories'. In *Concurrent Imaginaries, Postcolonial Worlds: Toward Revised Histories*, edited by D. Brydon, P. Forsgren, and G. M. Fur, pp. 33–57. Leiden and Boston, MA: Brill Rodopi, 2017.

Gallagher, John, and Ronald Robinson. 'The Imperialism of Free Trade'. *Economic History Review* 6, no. 1 (1953): pp. 1–15.

Giersch, Charles Patterson. *Asian Borderlands: The Transformation of Qing China's Yunnan Frontier*. Cambridge, MA and London: Harvard University Press, 2006.

Lee, James. 'Food Supply and Population Growth in Southwest China, 1250-1850'. *Journal of Asian Studies* 41, no. 4 (1982): pp. 711–746.

Ma, Jianxiong. 'Marriage and Land Property: Bilateral Non-Lineal Kinship and Communal Authority of the Lahu on the Southwest Yunnan Frontier, China'. *South East Asia Research* 19, no. 3 (2011): pp. 495–536.

Ma, Jianxiong. 'Shaping of the Yunnan-Burma Frontier by Secret Societies since the End of the 17th Century'. *Moussons N°* 17, no. 1 (2011): pp. 65–84.

Ma, Jianxiong. 'The Rise of Gentry Power on the China-Burma Frontier since the 1870s: The Case of the Peng Family in Mianning, Southwest Yunnan'. *International Journal of Asian Studies* 11, no. 1 (2014): pp. 25–51.

Ma, Jianxiong. 'Salt and Revenue in the Frontier Formation: State Mobilized Ethnic Politics in Yunnan-Burma Borderland since the 1720s'. *Modern Asian Studies* 48, no. 6 (2014): pp. 1637–1669.

Ma, Jianxiong. 'The Zhaozhou Bazi Society in Yunnan: Historical Process in the Bazi Basin Environmental System During the Ming Period (1368-1643)'. In *Environmental History in East Asia: Interdisciplinary Perspectives*, edited by Ts'ui-jung Liu, pp. 131–155. Abingdon and New York, NY: Routledge, 2014.

Ma, Jianxiong. 'The Dowry Land System and Chieftains of Shan-Dai Borderlands between Yunnan and Burmese Kingdoms from the Ming to the Qing Dynasties: The Construction of a Decentralized Institution in the Frontier'. In *The India-China Corridor workshop 'Between Yunnan and Bengal: Process geographies in the making of Modern Asia'*. Yunnan Minzu University, Kunming, China, 17–19 April 2018.

Mangrai, Sao Saimong. 'The Shan States and the British Annexation'. Ithaca, NY: Cornell University, 1965.

Myint-U, Thant. *The Making of Modern Burma*. New York, NY: Cambridge University Press, 2001.

Pollak, Olliver B. 'The Origins of the Second Anglo-Burmese War (1852-53)'. *Modern Asian Studies* 12, no. 3 (1978): pp. 483–502.

Pratt, Mary Louise. *Imperial Eyes: Travel Writing and Transculturation*. London and New York, NY: Routledge, 1992.

Sadan, Mandy. *Being and Becoming Kachin: Histories Beyond the State in the Borderworlds of Burma*. A British Academy Postdoctoral Fellowship Monograph. Oxford: Oxford University Press, 2013.

Saxer, Martin. 'Pathways: A Concept, Field Site and Methodological Approach to Study Remoteness and Connectivity'. *Himalaya – The Journal of the Association for Nepal and Himalayan Studies* 36, no. 2 (2016): pp. 104–119.

Thaung, Ma. 'British Interest in Trans-Burma Trade Routes to China, 1826-1876'. London: University of London, 1954.

'The Medical and Physical Society of Calcutta'. *Asiatic Journal and Monthly Register for British India and Its Dependencies* 25 (January–June 1828): pp. 225–226.

Van Schendel, Willem. *The Bengal Borderland: Beyond State and Nation in South Asia*. London: Anthem Press, 2004.

White, Richard. *The Middle Ground: Indians, Empires, and Republics in the Great Lakes Region, 1650-1815*. Cambridge Studies in North American Indian History. Cambridge and New York, NY: Cambridge University Press, 1991.

Willoughby, M. E. *Report on the Mules and Ponies of the Province of Yünnan (West China)*. Calcutta: Office of the Superintendent of Government Printing, India, 1904.

About the author

Gunnel Cederlöf, Professor of History at Linnaeus University, Sweden, and member of the Centre for Concurrences in Colonial and Postcolonial Studies. She studies environmental, legal, and colonial history in India and South Asia. Publications include *Founding an Empire on India's North-Eastern Frontiers, 1790-1840* (2014), *Landscapes and the Law* (2008) and *Ecological Nationalisms* (2006 with K. Sivaramakrishnan).

5 'Circulations' along the Indo-Burma Borderlands

Networks of Trade, Religion, and Identity

Joy L. K. Pachuau

Abstract

This chapter tries to understand the region designated as Northeast India in the context of trans-regional flows of commodities, ideas, and people. The argument is that such movements contributed to constituting the region, creating the identities of the various communities therein. Such an interpretation goes against the grain of popular and contemporary historiography, which largely views historical developments within isolationist frameworks. The circulation of cowries, horses, metals and, later, firearms suggests the wide networks of which Northeast India was a part. The intra- and inter-regional exchange of brides as well as religious ideas, besides commodities, also complicates the matter. The chapter thus explores the myriad ways in which the insularity and givenness of the region can be challenged.

Keywords: Northeast India, circulation, commodities, networks

This chapter is a step towards understanding the region that is today designated as Northeast India[1] in the context of trans-national networks that have contributed immensely to the making of the region. While the premise

1 The region designated as Northeast India includes the states of Arunachal Pradesh, Assam, Manipur, Meghalaya, Mizoram, Nagaland, Sikkim, and Tripura. The designation Northeast India received official status in 1971 when the Government of India created the Northeastern Council (NEC), a special government agency that plans and disburses funds for the social and economic development of the eight states. More recently, in 2001, a special Ministry of Development of Northeastern Region (MoDONER) was also created by the central government for a focused development of the Northeast.

Cederlöf, Gunnel, and Willem van Schendel (eds), *Flows and Frictions in Trans-Himalayan Spaces: Histories of Networking and Border Crossing*. Taylor & Francis Group 2022
DOI: 10.5117/9789463724371_CH05

may seem obvious enough, it is not so when we consider the historiography of the region. Firstly, Northeast India rarely figures in the historiography of India, and when it is studied it is usually problematized as a distinct region, except perhaps for the state of Assam (and sometimes Manipur) where deep cultural and religious linkages with the sub-continent are acknowledged at various periods in its history.[2] Secondly, the historiography of the region also privileges the constitutive state identities, especially in cases where so-called 'tribals' predominate.[3] Such histories often take state boundaries or communities, or both, as given. The studies are typically ethnicity or 'tribe' based, making the histories of the communities isolationist in their premises and understanding.[4]

What I therefore intend to do in this chapter is to see how this region was an important part of wider networks of trade, religion, commodity exchange, and the mobility of people. These networks extended in all directions, the limits of which cannot actually be drawn. The study thus investigates the extended regional connections that the Northeast was historically a part of. It also tries to map out the different kinds of connections that have existed within the region itself. If one were to place a compass in the Northeast, the reach of the pointer could be as far or as near as we want it to be – in other words, concentric rings of connections, albeit unstructured, and through intermediaries, could reach out far and wide through the centuries. Unfortunately, precise histories cannot be given, especially the exact dates and chronologies, as historical sources are few and far between. One of the difficulties in studying the region is the diversity of languages. More than 200 exist in the region and proficiency for a researcher in all of them is impossible. Added to this is the fact that written texts in most of these languages appear only with the arrival of colonial modernity, which itself did not make a uniform appearance in the region. There is therefore an over-dependence on the colonial archives, which despite their biases offer a wealth of information even as exploratory expeditions were able to

2 An important work in this regard is Sinha, *Tribal Polities*. In this volume, the contributors propose Hinduization as an important factor contributing to the rise of state systems (Ahom, Dimasa, Manipur, and even Jaintia) in the region, suggesting connections with the Indian sub-continent.

3 Northeast India has a large population designated as tribal by the Indian Constitution. There are 145 such communities in the region, 78 of which are considered major tribes with a population of over 5,000. The tribes constitute 27 per cent of the total population of the region, with some states, however, having more than 90 per cent of its population designated as tribal. See *Basic Statistics of North Eastern Region 2015*.

4 For example, Horam, *Social and Cultural Life*; Bareh, *History*.

identify the very many forms and modes of connections that existed in the region. By corroborating evidence from the colonial archives and, where possible, with the local sources, which includes the evidence provided by material culture, I have tried to draw certain conclusions about the reach of the region and the connections that the region was a part of. The evidence being patchy, one must also take various time periods into consideration. The question that follows from this is: what do these connections tell us about pre-colonial societies in Northeast India? How did such networks impact the societies that they were in? These are the issues that the study will address while highlighting some of the region's networks.

It must, however, also be admitted that notions of connections, especially those of trade and migrations and cross-cultural religious transactions, have not been totally absent from studies on the Northeast. For instance, several of the communities in Northeast India recognize a history of migration, especially from further east of themselves – although where exactly continues to be debated. Nagas, Mizos, and their kindred communities across their borders as well as some Khasi groups all claim migration from the east.[5] Communities that belong to the Tani group in Arunachal, while they claim migration from the north (Tibet) are, according to Stuart Blackburn, more likely to have also come from the east.[6] These movements, however, have usually been explained in local contexts and not as part of possible trans-regional or even global networks. The latter understanding has come about as a result of two kinds of historiographical shifts and interventions that have emerged in the relatively recent past, which enable us to see the region in a broader, more extensive perspective. The first is, of course, the idea of Zomia as proposed by Willem van Schendel.[7] This concept is important in that it posits Northeast India as part of a wider region of mountains – it gives the region a new spatial dimension not circumscribed by present political boundaries and enables us to see not only historical links but also the regional specificities shared with its coterminous regions. What Zomia establishes for us is the idea of a region and its interconnectedness based on a particular spatiality or landscape, which is an interesting and important way of understanding the past. This concept is taken further in this chapter when we see the seven states of the Northeast not as distinct and

5 Khasis claim migration from Southeast Asia, the Mizos from Southern China, and Nagas from Eastern Tibet, cf. Chaube, *Hill Politics*, pp. 1-2. Also see Lotha, *Articulating Naga Nationalism*; Pachuau, *Being Mizo*; Hutton, *Angami Nagas*, pp. 6-8.

6 Blackburn, *Himalayan Tribal Tales*, pp. 229-230.

7 van Schendel, 'Geographies', pp. 647-668.

separate entities, but as part of a larger region with connections across the present-day international boundaries, along its contiguous borders. Another important concept that also motivates this study is the idea of circulatory regimes in South Asia that was put forth by Markovits, Pouchepadass, and Subrahmanyam.[8] The authors tried to move away from a historiography that proposed a stable, immobile, and unchanging South Asia by showing the circulations and movements that occurred within and outside the region as factors of analysis rather than seeing them as independent phenomena that did not affect the socio-political economy of a region. Thus Markovits et al. had proposed the idea of circulation which focused not just on movements of people and goods, but also of information, knowledge, ideas, techniques, skills, cultural productions, religious practices, and even gods. To them, circulatory regimes suggested not only a movement one way, but back and forth, wherein 'men and things and notions transform themselves' when they move. It also concerns the totality of circulations that occur in a given society.[9]

What both these interventions offer us is the means to look at Northeast India, not in isolation but as intrinsic to and part of larger global networks. It also enables us to see that the social processes that occur in the region are not merely the result of internal developments but that they evolved in relation to larger processes elsewhere.

This chapter would be too ambitious if it tried to uncover all the different kinds of connections that the Northeast was a part of. However, it is a tentative step in that direction. I begin by first placing the Northeast within the scope of wider networks from beyond the region that have existed historically to see how this region has been involved in global processes. Herein I expand the work of Bin Yang[10], who explores the networks of silver, horses, and cowries in the Southwest Silk Route (SSR) to see whether these routes and commodities encompassed the Northeast in the pre-colonial period. Firearms are another commodity with a presence that is noted in the immediate pre-colonial period; they provide an interesting example of the wide reach of the commodity network of which the Northeast came to be a part of. I then look at intra-regional networks focusing on religion, courtly exchanges, and traders to see the kinds of associations that have made the region. In doing so, we see the different network regimes in operation; we also see how the mountain fastnesses, often seen in isolation as well as isolationist, formed a region of interconnectedness.

8 Markovits et al., *Society*.

9 Markovits et al., *Society*.

10 Yang, 'Horses', pp. 281-322.

The region under consideration

The region east of Bangladesh that we know today as Northeast India is a region that is crisscrossed by mountains, be they the Himalayas to the north (encompassing present-day Sikkim and Arunachal Pradesh with Bhutan in between, also known as the Eastern Himalayas); the Karbi-Meghalaya plateau with the Khasi, Jaintia, and Garo Hills; or the mountains of the great Indo-Burman ranges, stretching from Nagaland to the Rakhine state of Myanmar and incorporating the Naga Hills, the Somra tract, the Lushai (Mizo) Hills, the Chin Hills, the Chittagong, and the Arakan Hill tracts. The mountain ranges extend further east into China as far as Sichuan, as well as into Burma as far as the Chindwin and Irrawaddy river valleys. There is thus a mountainous specificity to the region, although these mountain areas in Northeast India are also surrounded or interspersed by river valleys such as the Brahmaputra, Barak, and Manipur river valleys. These valleys and plains have been the areas where the more established state systems have emerged. In the period preceding colonial conquest, the ruling dynasties that had established control were the Dimasas and Ahoms in Assam, the Ningthoujas in Manipur, and the Manikya rulers in Tripura. There were also more established and settled hill chieftaincies like that of the Khasi and Jaintia rajas of present-day Meghalaya.[11] Similarly, in Myanmar, to the east of the hilly tracts of the region, the Irrawaddy river valley has seen different dynasties establish their domain, and in the period preceding the British, it was the Konbaung dynasty that had established itself in the eighteenth century, first at Ava and then at Mandalay. To the west of the Northeast, the Bengal plains had regional stable state systems from the eighth century. In Yunnan, too, several powerful dynasties were in control until they were brought within the ambit of China under the Yuan dynasty in the thirteenth century. It is therefore useful to see these mountain regions with their chieftaincies or village republics in relation to the states that had developed in the region – especially in the valleys. Contiguous mountain ranges and their inhabitants were interspersed with and surrounded by state systems of varying prominence and formation, and all of them had varying levels of interaction amongst themselves as well as with the hill states/tribes. The inhabitants of the Northeast had also migrated into the region and were therefore not autochthonous. The Ahoms, for example, were Shans who migrated from upper Burma in the thirteenth century and established their own kingdom of the same name, initially in the upper reaches of the river

11 Sinha, 'Introduction'.

Brahmaputra in Assam. Many such examples of movements and migrations exist in the region. This spatial context is important for us to know, as we will see that neither the hills nor the valley states existed in isolation. The existence of powerful state systems that surrounded or interspersed the otherwise rugged terrain of hill chieftaincies suggests the possible push and pull factors that facilitated the formation of networks of connections. For instance, Angma Jhala's important work on the Chittagong Hill tracts shows the myriad ways in which the zones that adjoin the Bengal plains and the hills east of them developed a rather complicated landscape of religious, political, and ethnic mixing even as they lay in the midst of many political (Mughal, Arakan, Burmese, and Ahom as well as Portuguese) and ecological zones.[12] In fact, she goes on to argue that the complex structure of society that had developed there ensured that the classificatory systems employed by the British often found the region with its many syncretistic religious, political, and linguistic practices difficult to categorize. What the work therefore suggests is a history of intense borrowings across ecological and political zones.[13]

As a caveat, I must also state here that 'Northeast' was a term that was invented in colonial times with changing understandings of its extent throughout the colonial period. In 1816, David Scott, an English East India Company official, had been appointed judge and magistrate of Rangpur, a northeastern district in Bengal. With the commercial aims of the Company increasing and the need to establish zones of influence, David Scott was further made Governor General's Agent to the North East Frontier with 'general control and superintendence and intercourse with the petty states of Sikkim, Bhutan, Tibet, Cooch Behar, Bijni, Assam, Cachar Manipur, and Jaintia.'[14] In the course of time, some of these states were dropped from what came to be known as Northeast India. While such a regional understanding was not prevalent in the pre-colonial period, I address the region as Northeast even in the pre-colonial period to give a territorial specificity to the region.

Long-distance trade networks and Northeast India

In their work on maize in Northeast India, Stonor and Anderson had speculated for a pre-Columbian entry of some varieties of corn into the

12 Sinha, 'Introduction'.

13 Jhala, *Endangered History*.

14 Chaube, *Hill Politics*, pp. 4-5.

region from the Americas. According to them, certain varieties such as popcorns, green corns, and waxy corns resemble those that were grown in Peru and Chile in pre-historic times.[15] While their work highlights the possibly deep global connections with regard to the dispersion of grains, it is not possible at this stage to take their arguments further due to a lack of evidence. However, works about the Southwest Silk Route (SSR) enable us to recognize a little more explicitly the networks of trade that existed in the region surrounding the Northeast amongst more established state systems. According to Bin Yang, the SSR had a maritime as well as an overland route, with the overland route passing through Yunnan, Upper Burma onwards to India through Assam.[16] He traces the history of three commodities: cowries, horses, and silver, to suggest the wide networks that this route encompassed. Although the history of the SSR goes as far back as to the second century BCE, concrete evidence such as the routes and the volume of exchange, especially with regard to the Northeast and for the entire time span since then is vague. However, what is definite is that the movement of all three of these commodities also affected the Northeast. We know, for instance, that cowries were acquired from the Maldives in the Indian Ocean, from where they were taken and used in the Bay of Bengal region, mostly as petty currency. The final destination of these cowries was Yunnan. Interestingly, cowries did not have the same currency beyond Siam or Thailand, or beyond Yunnan for that matter. Similarly, as regards horses, there is evidence to show that the horses of Yunnan were extremely prized creatures,[17] exported far and wide, with Bengal rulers acquiring their horses from Yunnan, too. With evidence from Taranatha's *History of Buddhism in India and Tibet* and Minhaj-Us Siraj's *Tabaqat i Nasiri* (both thirteenth-century works), M.N. Rajesh comes to the conclusion that at least 1,500 horses came to the plains of Bengal from Yunnan and Tibet every day.[18] The same is the case with silver. Silver bullion was shipped from Yunnan and Upper Burma into Bengal between the thirteenth and sixteenth centuries.[19]

Bin Yang's assertions are in the context of placing the SSR within a global economy, but the question that is pertinent for us is whether the Northeast played a role in or was part of these trade networks. Considering the fact

15 Stonor and Anderson, 'Maize', pp. 355-404.

16 Yang, 'Horses', pp. 281-322; Yang, 'Rise and fall', pp. 1-25; Yang, 'Bay of Bengal', pp. 317-342.

17 Jianxiong Ma and Cunzhao Ma inform us that although the horses of Yunnan were small, they were appreciated for their strength. They were also prized as warhorses. Cf. Ma and Ma, 'Mule caravans', pp. 24-42.

18 Rajesh, 'Tibet'.

19 Also see Hussain, 'Silver', pp. 264-308.

that the evidence for the circulation of these commodities in and around the Northeast is clear, we therefore need to see if these commodities entered the region and whether the region was part of these better-known and well-established connections. And, although evidence is patchy, especially as we delve deeper into the past, the Northeast was not isolated from these exchanges. We will see that commodities that connected Yunnan with Bengal, namely silver, horses, and cowries, also made their way to the Northeast.

Let us first turn to the evidence for *silver* in the Northeast in the pre-colonial period. The Ahoms were known to have struck their own coins in gold and silver from the sixteenth century.[20] According to colonial officials writing in the early nineteenth century, silver came to Assam from China or Tibet.[21] F.C. Henniker, a colonial official writing in the early twentieth century, also reported the traditional use of silver in Assam, Manipur, and the Khasi hills, albeit more as jewellery rather than currency.[22] However, most of this silver was said to have come from Bengal. Bengali and Manipuri artisans were employed by the three communities (Assamese, Meitei, and Khasi) to make their jewellery, although – according to Henniker – the designs were 'national', meaning local or indigenous. The use of certain people as artisans suggests the circulation and movement of people in the cause of this trade. Henniker goes on to say that the use of silver was, however, limited, even as many other places such as Darrang, Nowgong (both in Assam), the Naga hills, and Lushai hills did not know its use.[23] On the other hand, J.P. Mills noted the popular use of silver jewellery amongst the Mishmis (of present-day Arunachal), which he said was made from Yunnan silver coins.[24] Scattered evidence thus shows that the Northeast was taking advantage of the much more established silver connection between Yunnan and Bengal and had access to it. Although the location of the Northeast between Yunnan and Bengal would have made it an ideal conduit connecting the two regions, there is no evidence of this. However, it is clear that the silver in the Northeast was acquired directly from Yunnan and Bengal, and that the areas adjoining them had more access to it.

We now turn to *horses* or ponies in the region. The royal court chronicle of Manipur, the *Cheitharon Kumpapa,* mentions that polo was played for the first

20 Allan, 'Coinage', p. 301.
21 M'Cosh, *Topography*, pp. 58, 66.
22 Henniker, *Gold*.
23 Henniker, *Gold*.
24 Mills, 'Mishmis', p. 3.

Figure 5.1 Items of Khasi jewellery in silver[25]

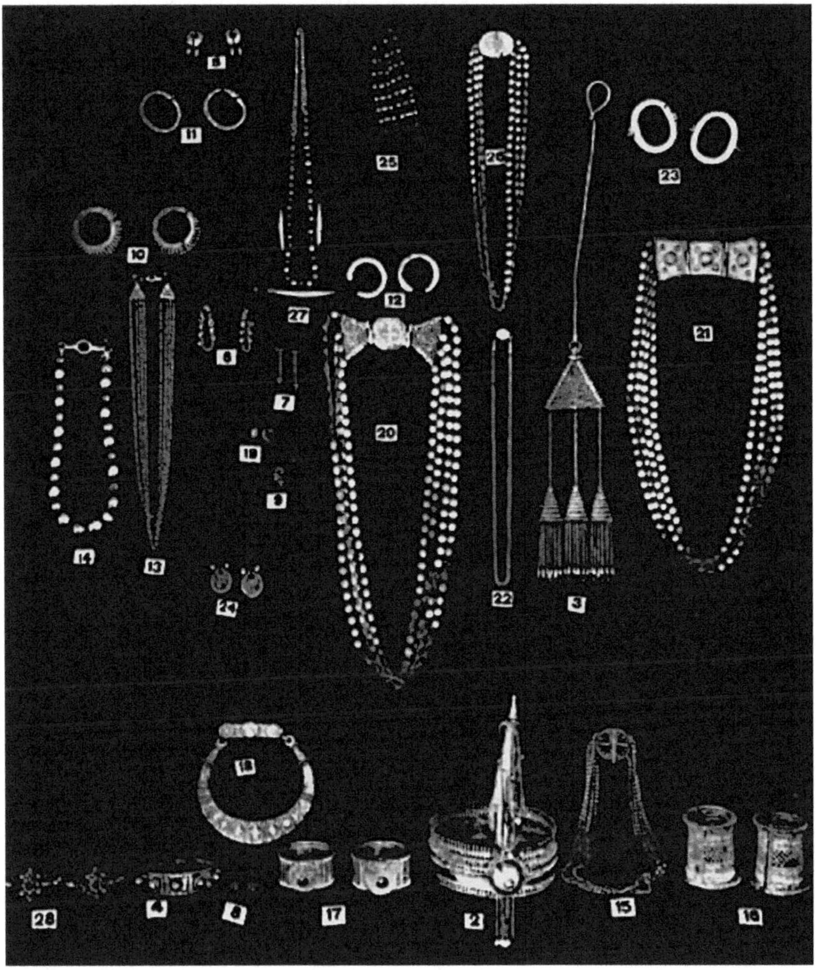

time in the early seventeenth century, suggesting the presence of horses from quite an early period.[26] In the nineteenth century, R.G. Woodthorpe, a colonial official, spoke about the Manipuri horses as 'active, hardy little ponies' that were being sold by Manipuri horse dealers to them during the Lushai expedition of 1871. He also mentions the game of 'hockey', which was played using these horses.[27] Today, the sport is played on what are called Manipuri ponies that

25 Henniker, *Gold*, p. xxxi
26 Parratt, *Court Chronicle*, p. 70.
27 Woodthorpe, *Lushai Expedition*, pp. 56-57.

are known to be 'extremely agile and tough' having great 'stamina, speed, and intelligence'.[28] However, it is also acknowledged that the ponies are not indigenous to the region and that they originated from Mongolian wild horses that were crossed with other breeds.[29] There are scattered pieces of evidence from the past to suggest that they came from further east but also from the Bengal plains. The *Cheitharon* speaks of two horse breeds in Manipur: the Mayang *sa* (horses from Cachar and Sylhet) and the Mangkan *sa* (other sources: Mangal (Mongol) *sa*) suggesting the places from where they were acquired.[30] The same chronicle also mentions that groomsmen for horses were captured from Bengal (Pangal), suggesting the connections that Manipur had with Bengal.[31] There is evidence to show that in the fifteenth century, King Kyamba of Manipur took ponies from Burma as spoils of war.[32] The *Cheitharon* also mentions that horses were acquired from Aawa (Ava/Burma) in the eighteenth century.[33] Thus, polo, a game known in Manipur even prior to the arrival of the British, became possible when horses, which were not indigenous, were imported.

Regarding horses in the Ahom kingdom, it was said that the first Ahom ruler, a Tai prince, entered Assam in the thirteenth century with 3,000 horses across the Patkai passes from Upper Burma. However, the Ahoms seem to have subsequently given up horses, possibly because they found elephants more effective in warfare and perhaps also because their supply across the Patkai had ended.[34]

There are other scattered pieces of evidence of the presence of horses from the hill communities in the period preceding colonial occupation. Carey and Tuck in their work relating to the Chin Hills (Myanmar) noted that the chief of the Sokte tribe in the late nineteenth century had ridden a pony in his campaigns.[35] This is quite a rarity because we seldom have evidence of the use of ponies amongst the hill tribes. A petroglyph on a memorial stone platform (megalith) in Mizoram, a region parallel to the Chin Hills but west of it, known as *Lalpuithanga lungdawh* (in Serchhip), said to be from the late eighteenth century also depicts a horse, a very uncommon feature for petroglyphs from the region.[36]

28 Manipuri Pony, http://imh.org/exhibits/online/breeds-of-the-world/asia/manipuri-pony/index.html, accessed 24 October 2018.

29 Manipuri Pony, http://imh.org/exhibits/online/breeds-of-the-world/asia/manipuri-pony/index.html, accessed 24 October 2018.

30 Parratt, *Court Chronicle*, pp. 179-180, Glossary.

31 Parratt, *Court Chronicle*, p. 104.

32 Manipuri Pony, http://imh.org/exhibits/online/breeds-of-the-world/asia/manipuri-pony/index.html, accessed 24 October 2018.

33 Parratt, *Court Chronicle*, p. 127.

34 Hazarika, 'Study', pp.17-22; Ahmed, 'Elephants'.

35 Carey and Newman Tuck, *Chin Hills*, p. 118.

36 Malsawmliana, 'Megalithic monuments', pp. 52-62; Laldinpuii, 'Memorial stone'.

Figure 5.2 Memorial stone (*lungdawh*) of Lalpuithanga, 18th century[37]

37 According to K. Laldinpuii, the stone platform was erected to honour Lalpuithanga, a Sailo chief of the Lushai Hills (Mizoram) probably in and around 1886CE after the first British military expedition to the Lushai hills during 1871-2. The large human figure in the middle was probably the chief Lalpuithanga himself. Along with the central figure, several animals either killed or owned by the chief are depicted to suggest that he had become a *thangchhuah* or a 'distinguished man'. The figures include those of a slain man, a monitor lizard, several birds, pigs, elephants, tigers, and the horns of a mithun. There are engravings of two horses on either side of the stone, which suggests that Lalpuithanga may have owned horses (mizomegaliths.blogspot.in, accessed 15 May 2012).

With regard to where the horses could have come from, Carey and Tuck mention the use of 'Panthay mules and drivers' on the tracks between the Lushai (Mizo) and Chin Hills. Panthay was the Burmese word that referred to Chinese Muslims, more commonly also referred to as Hui, who had come to Burma for trade from Yunnan and had settled in upper Burma, especially around the Shan area. The Shan area was traditionally the area from where mules were obtained for regions to the west of it.[38] Jianxiong Ma's work also endorses this understanding even as he talks of a very active caravan trade mainly carried out by Hui merchants between Yunnan and Burma from a very early period.[39]

Like silver, horses in the Northeast were thus acquired either from Bengal or from Burma. Ultimately, however, all the horses including those in Burma and Bengal were acquired from Yunnan. Horses were greatly prized in the Manipuri court with special groomsmen attending them, and even amongst the hill chieftains, they were becoming an important item of status and prestige, rare as they were.

The other commodity that was important in the long-distance exchange between the Maldives and Yunnan, which also had the Northeast in its ambit, was cowries. Here, too, a more thorough investigation is required, but a summarizing enquiry reveals the widespread use of cowries in many parts of the Northeast, although certain communities were not within the exchange networks of this commodity. Thus, for instance, Gurdon has noted the Khasi use of cowries as a medium of exchange and payment for labour, besides its use in settling divorces, divination, religious rituals in certain types of deaths, and in certain games played by children.[40] Endle writes that among the Kacharis (a community in present-day Assam), a class of priests called *ojhas* 'manipulated' shells and cowries to predict the future.[41] The Lai and Mara communities of southern Mizoram used cowries in attire as items of prestige, and only women from the chief's household were entitled use them in their *puans* (sarongs).[42] Conversely, the Mizos in northern Mizoram do not seem to have been exposed to the use of cowries. Many of the tribes in Nagaland as well as Arunachal are also known to have used cowries and conch shells as items of prestige and status markers.[43] Thus,

38 Chang, *Beyond Borders*.

39 Ma and Ma, 'Mule caravans', pp. 24-42.

40 Gurdon, *Khasis*, pp. 57, 64, 80, 117, 119, 136.

41 Endle, *Kacharis*, p. 40.

42 Parry, *Lakhers*, p. 38.

43 Blackburn, 'Memories', pp. 15-60. Blackburn's essay provides an interesting window into the trade networks of the Himalayan region and Arunachal Pradesh's place in that trade.

while in most of the coastal societies such as Bengal, cowries were used as a medium of exchange, as petty currency,[44] in other parts of the Northeast they were transformed into status items.

I believe that although cowries are found in places in the Northeast that are nowhere near coastal areas, their prevalence does not show routes of migrations as some are wont to believe,[45] but participation or inclusion in trading networks that stretched from the Maldives to Yunnan via Bengal. According to Bin Yang, Assam had a monetary system based on the cowrie before the fifteenth century, while in Arakan and Pegu white cowries were used as coinage.[46] Mir Jumla, the Mughal governor, noted the use of gold and silver in Assam as coinage; however, he was surprised to note that for petty currency, cowries were used, not copper.[47] According to David Ludden, Bengal used cowries as coins quite extensively up to the end of the eighteenth century, but in places like Sylhet the trading was almost totally dependent on cowries. The people of Sylhet brought almost nothing except cowries from their downstream trade with the plains of Bengal in return for which they sent rice, fish, and other upland products. These cowries originated in the Maldives, from where they were transported to Chittagong or Calcutta (now: Kolkata), stored in Dacca (now: Dhaka), and brought upstream to Sylhet. The geographical extent of cowrie country, according to Ludden, was 'the spatial dispersion of markets that connected uplands and lowlands around the Sylhet basin.[48] It is therefore possible that cowries were exported from Sylhet to other places in the Northeast, especially its western borders that were contiguous with Bengal. The other route for the entry of cowries into the Northeast was through Burma, most probably the Hukawng valley. This is consistent with its importance among the Nagas and the eastern Arunachal tribes. According to Vogel and Hieronymus, the cowries that Yunnan imported came via land routes through Burma and

44 The exchange rate for cowries in Bengal in the eighteenth century. '4 cowries make a gonda, 5 gondas a budi, 4 budis a pan, 16 pans a kahan, and 10 kahans a rupee (a silver coin of approximately 172 grains). Another variant of the same counting system was: 4 cowries make a gonda, 20 gondas a pan, 4 pans an anna, 4 annas a kahan, and 4 kahans a rupee (28,880 cowries).' 'Cowrie', http://en.banglapedia.org/index.php?title=Cowrie, accessed 4 August 2021.

45 Shimray, *Origin*, p. 13, as cited in https://thohepou.wordpress.com/2007/08/17/the-myths-of-naga-origin/. According to this theory, the Nagas belong to a group that migrated southwards from China and made their way back after reaching the southern seas near Moulmein/Mawlamyine (in Myanmar). Others believe the migration could have come from Indonesia, the Philippines, or Malaysia.

46 Yang, 'Laws', p. 165.

47 Allan, 'Coinage', pp. 300-331, p. 311.

48 Ludden, 'First boundary', pp. 1-54.

Siam, the cowry here sometimes being imported from the Philippines and the Moluccas.[49] As we will see, trade routes between the Naga regions and the Hukawng valley were not unknown in pre-colonial periods.

How far back this participation took place cannot be determined, but as was pointed out by J. H. Hutton, in the early twentieth century, Angami traders from Nagaland would go to Calcutta or Dacca to acquire conch shells,[50] whilst cowries were brought from Myanmar to Tuensang and Khonoma in the Angami region. According to Kanungo, cowrie hoards have also been found on archaeological sites in Assam, and it is possible that they were either used as currency or as items to be traded with the Nagas.[51] According to Kanungo, S.N. Majumdar had in 1924 noted that the Naga exports 'for sale in the plains were cotton, rubber, tea, sugarcane and cardamom, while the chief imports were beads, bangles of glass, zinc or brass, boar tusks, ivory and glass. These were chiefly imported from Kolkata, Mumbai and Myanmar. He further noted that the Angamis were engaged in widespread commercial activity and used a public path to Myanmar through Manipur. For this travel, the Sema hills route was generally adopted.'[52] Similarly, although Gurdon does not mention where the Khasis obtained their cowries from, he writes that the Khasis obtained coral beads from Calcutta.[53]

The routes mentioned above constitute evidence from the twentieth century. However, there is no reason to believe that these routes were established due to the arrival of *pax Britannica*. In fact, the paths seem to have been age-old routes that the colonial writers took notice of, and it is perhaps safe to say that colonialism instead put an end to many of these routes through the creation of political boundaries – the Inner Lines being one such means of doing so.[54] In the mid-eighteenth century, a colonial official reported about a route that existed between present-day eastern Arunachal and Yunnan. According to him, 'the inhabitants of a Singpho village at the mouth of the Terap were in constant dealings with the Hookong [Hukawng] valley of upper Burmah, the locale of amber mines and petroleum springs in that country ...'.[55] It was also reported that a man could take eight days

49 Vogel and Hieronymus, 'Cowry', pp. 211-252, in particular pp. 237-245.

50 Hutton, 'Naga Chank', p. 223; Kanungo, 'Naga ornaments', p. 160.

51 Kanungo, 'Naga Ornaments', p. 161.

52 Kanungo, 'Naga Ornaments', p. 160. Also see Allen, *Assam*.

53 Gurdon, *Khasis*, p. 22.

54 Pachuau and van Schendel, 'Borderland Histories', pp. 1-4. The Inner Line was instituted in 1873 with the aim of prohibiting people outside the area from entering the region. However, it also had the effect of preventing the people outside the Line from entering British territories.

55 Goodenough, 'Letter', p. 334.

to reach Hookong without a load. The official reported that two routes led from there to China, one in a southerly direction via Bhamo and the other in an easterly direction across the Irrawaddy cutting through the mountain ranges that separated China from Burma. Furthermore, the document mentioned that the Chinese also ventured up to Hukawng either via Bhamo or on other more direct routes.[56] Similarly, according to Pommaret, 'Assamese merchants went to Yunnan in China by the line of trade through Sadiya, Bisa and across the Patkoi range of mountains, and through the Hukawng valley to the town of Munkong from where they ascended by the Irrawaddy to a place called Catmow. The goods were disembarked at Catmow from where they were conveyed on mules over a range of mountainous country inhabited by the Shans into the Chinese province of Yunnan.'[57] The route was said to be in existence from the seventh to the eighteenth centuries. Although such routes did not develop to be crucial commercial axes, it is clear that a certain amount of merchandise was exchanged between Ahom and Yunnan through Burma with the Burmese Singpho community playing an important part.[58]

In the period immediately preceding the arrival of the British, we also find another commodity that makes its appearance in Northeast India. I am referring here to firearms. Their appearance also suggests long-distance networks, albeit through intermediaries.

Let us now digress a little and turn to the history of firearms in order to see how the region came to be part of this global exchange. Even though the Songs (960-1279) had used incendiaries in Taoist rituals, the development of small firearms had begun in the west. Major strides were made in the eighteenth and nineteenth centuries with improvements in technology – the period is often referred to as the breech-loading revolution. Gun technology had moved from muzzle-loading flintlocks or muskets that had first been made in the seventeenth century (a famous one being the smooth-bore flintlock known as the Brown Bess musket made in the early eighteenth century) to that of rifled breech loaders, which made them more efficient. According to Emrys Chew, while the big empires in Asia – Turkey, Persia, and India – concentrated on siege artillery, the smaller states in Southeast Asia such as Burma, Siam, and Vietnam were quick to adopt and adapt the smaller handguns and light weapons, which no doubt played a role in the

56 Ibid.

57 Pommaret, *Ancient Trade Partners*, https://www.repository.cam.ac.uk/bitstream/handle/1810/227005/JBS_02_01_02.pdf?sequence=2.

58 McCosh, 'Account'.

political equations of the region. It was Europe that had transferred the use to small portable firearms, and the earliest of these examples came to India and Southeast Asia through the European trading companies. While the European companies did not participate in the arms trade directly, they used it as an instrument of statecraft when firearms were transferred as a means of gaining concessions. Southeast Asian states also acquired such weapons when European soldiers were captured in times of war and their arms were confiscated; or as in the case of the Portuguese when renegades were employed by various courts of Southeast Asia. Technological changes in firearms manufacture also provided for large numbers of gun transfers even as the old ones became obsolete for the Europeans, and these guns found their way as smuggled goods to various parts of the world.[59]

Such guns found their way into the Lushai-Chin Hills, and it was said that almost all tribes had access to them.[60] According to L. W. Shakespear, the expeditions of 1871 and 1872 into the Lushai Hills revealed that the Lushais were the best armed in comparison to most of the tribes in the Northeast. In fact, according to Dzüvichü, it had been reported that the 'Kukis' of the Chittagong Hill tracts had possessed firearms since 1863, and that the Arakan had become a hotbed for the acquisition and supply of arms to the hills. Furthermore, in 1869 a report mentioned that all the tribes from the Garo and Jaintia Hills through to the Cachar Hills and southwards to Chittagong were in possession of firearms made in Birmingham that came via Calcutta and Backergunge (the southernmost district of the Dacca Division in colonial India).[61] Their firearms mainly comprised old percussion or flintlock muskets. Many of them were of English makes, Tower marked, others with the name Alton stamped on them.[62] According to Carey and Tuck, the firearms found in the Chin Hills also included some with French marks and stamps. Occasionally there were Russian, German, and American weapons, too, but 98 per cent of them were said to be English makes.[63] It was clear that these guns that were acquired came from the east and the west of the Chin-Lushai Hills. Some of the weapons had names of Burmans in Burmese characters, while the others had names of Indian sepoys in Persian characters 'stamped on the heel-plates of the muskets.'[64] Some of the muskets

59 Chew, *Arming*; also see Carey and Tuck, *Chin Hills*, p. 223

60 British Library L/PS/6/547, T.H. Lewin, Dist. Supdt, CHT to the IGP, Lower Province, 17 Feb 1866, pp. 39-40.

61 Dzüvichü, 'Colonial State', pp. 226-227.

62 Shakespear, *History*.

63 Carey and Tuck, *Chin Hills*, p. 221.

64 Carey and Tuck, *Chin Hills*, p. 222.

could also have been captured from British soldiers as they had regiment numbers on them.[65] Most guns for the Lushais came from the Chittagong Hill Tracts (CHT), through Rangamati and Kassalong. According to reports, a well-to-do merchant of unknown nationality had made a lakh of rupees by selling firearms to the tribes north of the Koladyne.[66] By 1860, the sale of arms and ammunitions had been prohibited in the Chittagong district, but there were frequent reports of them being smuggled into the hills. Lewin, a British official in the CHT, reported in 1866 that Magh traders were secretly selling them to the hill tribes at the Poang *haat*s of the Chittagong district, and on one occasion 30 muskets and 300 rupees worth of powder had been smuggled in this way.[67] After the Chin and Lushai Hills had been successfully brought under British control in the 1890s, the secret re-arming of the Chin Hills took place through arms traded via the un-administered territory between the Chin, Lushai, and Arakan Hill tracts, suggesting the resilience of these trade networks despite the disarming and licensing of guns (begun in the Lushai Hills in 1894-5).[68]

According to Bertram Carey, the Chins and Lushais had also acquired firearms as early as in the 1830s – from Burma. There was an increase in the numbers of guns in the region with the annexation of Burma in 1885, as disarmament was carried out among the Burmese, and these guns found their way to the hill tribes. The Burmese preferred to sell them to the Chins rather than hand them over to the colonial government.[69] It was not just disarmament but also the introduction of new technologies of firearms that led to the dissemination of the old obsolete guns. As L. W. Shakespear noted, 'the introduction of breech loading rifles and the consequent discarding of old-time firearms which were brought up by the Native states and traders, diffused at first cast numbers throughout India, many finding their way into the hands of border tribes.'[70]

There was also a regular trade in these weapons from the early 1870s from Burma, Bengal, and Cachar, so that by the end of the decade, the Lushais, Kukis, and Angamis were in possession of considerable numbers, with many of them making their own modifications. According to some British officials, the Lushais had made these guns their own by painting and varnishing 'the

65 Carey and Tuck, *Chin Hills*, p. 222.

66 Rosanga, 'Consolidation', p. 99.

67 British Library, L/PS/6/547, T.H. Lewin, DIst Supdt, CHT to the IGP, Lower Province, 17 Feb. 1866, p. 21.

68 Rosanga, 'Consolidation', pp. 101-102.

69 Carey and Tuck, *Chin Hills*, p. 220.

70 Shakespear, *History*, p.78

musket stocks and butts in red, black, and yellow.' They also had '... their powder flasks of *gayal* (the mithun) horn [which] are beautifully polished and inlaid with silver or ivory.'[71] Carey also remarked that the Chins hated weight in the gun and therefore discarded the stock of western manufacture and carved one of their own pattern and unique to themselves from willow or a species of beech; this custom was universal throughout the hills.[72]

Another aspect of the spread of firearms that colonial officials often commented upon was the local manufacture of gunpowder. According to Carey, the people of the Chin Hills had initially obtained the powder from Burma as there was very little sulphur in the hills. But once its export was prohibited (1885), ingenious methods emerged. The chemical was obtained by soaking and boiling the sulphur-rich 'aunglauk' bean. Saltpetre and nitre were obtained from manure collected in baskets and strongly impregnated with urine. The liquid was drained into receptacles, boiled, and when it evaporated, crystals of saltpetre and nitre were produced. On the other hand, L. W. Shakespear speculated that the Nagas and the other tribes may have obtained the knowledge for making gunpowder from the Ahoms who were making it themselves in the seventeenth century. Mir Jumla[73] had commented upon their use in cannons in that same century, while Tavernier also alluded to the Ahoms' knowledge of making gunpowder, which may have been learned from the Chinese or the Portuguese, or through earlier struggles with the Muslim rulers of Bengal. The circulation of firearms thus suggests that not only was there a circulation of the commodity, but also a circulation of knowledge with regard to the local production of gunpowder.[74]

An indication of the extent of the circulation can be gauged from the numbers that were confiscated when British officials entered the region. For instance, Carey reported that on 1 April 1894, the gun register had shown that the Chin Hills had been disarmed of 4,302 guns. A single tribe, the 'Klangklang', was practically totally disarmed, in that 490 guns had been taken from the tribe, which contained about 975 houses, i.e., one gun for every two houses.[75]

The widespread use of firearms thus shows a connection that brought the tribes of the hills into contact with the valley states through a commodity that was important for statecraft. The arrival of the firearm brought about

71 Shakespear, *History*, p. 77.

72 Carey and Tuck, *Chin Hills*, p. 224

73 A seventeenth-century Mughal governor of Bengal who attacked the Ahoms.

74 Shakespear, *History*, p. 78.

75 Carey and Tuck, *Chin Hills*, p. 116

many changes into the political economy of the hill tribes, one of them being the role they played in the story of migrations from the east to the west of many of the communities. It is an acknowledged fact that the populations of Burma were the result of migrations of large groups such as the Mon-Khmer, the Burmese, and the Shan over several centuries.[76] At the time of the British arrival into the region in the nineteenth century, they were also witness to these migrations, especially in and around the regions they designated as Kuki, Chin, and Lushai areas (present-day south Manipur, Chin Hills of Burma, and Mizoram). According to J. Shakespear, the colonial official in the Chin and Lushai Hills, it was quite clear that it was the British who had put an end to the constant state of migrations of the pre-colonial phase.[77] Many reasons have been given for these migrations, from the practice of *jhumming*/swidden cultivation, which required the need to look for new territories every few years,[78] to a population explosion.[79] A much lesser known reason that was put forth by Pearn[80] and later Stevenson was the acquisition of guns. Pearn pointed out that it was the capture of French guns and gunners at Syriam in the eighteenth century that had contributed to the expansion of the Burmese state. Similarly, Stevenson notes:

> ... the Great Kuki invasion of Tipperah in Assam in 1860 and the westward migrations of the Lushais and other tribes in the belt of mountain country between Burma and Assam arose out of the acquisition of flint-lock guns by the Haka tribes of the central Chin Hills [which enabled, I believe, the beginnings of the processes of state-making]. The new weapons upset the Lushais out of their ancestral lands and thus start the latter on the predatory land and head-hunting expeditions to the west and north, which eventually led them into contact with the plains of India and brought them under British administration.[81]

Thus, the struggle for supremacy of tribes and villages in the region seen in the constant warring amongst them as witnessed by colonial officials was considerably facilitated by the introduction of firearms.

All the commodities mentioned above – silver, horses, cowries, and firearms – and their availability in Northeast India suggest that the region

76 Stevenson, *Hill Peoples*, p. 2.
77 Shakespear, *Lushei Kuki*.
78 Ibid.
79 Lehman, *Structure*.
80 Pearn, *Burma*.
81 Stevenson, *Hill Peoples*, p. 2.

was not isolated from the larger circulation and flows of commodities in and around it. The well-known trade routes of Yunnan-Tibet-Bengal or Yunnan-Burma-Bengal surrounded the Northeast, and smaller networks clearly infiltrated the region as well. Each commodity, however, also gained new uses and meanings: silver was transformed in its use from currency to jewellery, cowries from monetary unit to items of status or instruments of divination, horses as pack animals to their use in sport to symbols of status and royalty. Firearms, while being instruments of status, were also able to completely alter the polities of the region. New symbolisms and attachments were thus made using these products that came from worlds and networks beyond them.

Networks closer to home

As I discussed earlier, we can understand interconnections in and around Northeast India as concentric rings of relationships. However, smaller connections were also prevalent, and this is what we will now focus on. Here, Markovits's idea of circulatory regimes is particularly helpful. I focus on three such regimes, namely connections in the areas of religion, the exchange of courtly culture including marital alliances, and the movement of goods and traders.

One of the areas that we find the Northeast has always been receptive to was the area of religion. All the states mentioned earlier had emerged in the contexts of tribal consolidation and the adoption of one of the so-called 'world religions'. This establishes the fact that religion offered great scope for facilitating networks of connections.

According to Amalendu Guha, several tribal formations in Northeast India crystallized into 'rudimentary state formations' by the fifteenth century. There were many causes for this including the adoption of wet rice cultivation, the introduction of the plough, the ability to subjugate neighbouring populations, and the ability to involve themselves in minor commodity production. Another important cause, according to Guha, was the Hinduization in varying degrees of the ruling families.[82] The Tai-Ahoms, whose entry into Assam dates to the thirteenth century, began to have Brahmin influences in court as of the fourteenth century. They further consolidated their rule by collaborating with the sixteenth-century popular movement of Shankardeva (also known as Neo-Vaishnavism) which was

82 Guha, 'Ahom'.

followed by the lower classes, even though the Ahoms themselves were worshippers of the Shakti cult.[83] The Manikya rulers of Tripura had a similar story with the rulers adopting Hinduism by the fifteenth century, while the Dimasas came under Hindu influence in the sixteenth. Similarly, the kingdom of Manipur had strengthened itself further when the Ningthoujas adopted Vaishnavism in the eighteenth century.

Buddhism also made its mark in the region. Arakan, which was at the southern end of the mountain ranges, being at the borders of Burma and Bengal (a Mughal frontier), had Buddhist rulers who had Muslim names.[84] The Konbaung dynasty that ruled in Burma on the eve of British rule was decidedly Buddhist, but Burma had made the shift from Mahayana to Theravada (Hinayana) in the eleventh century under the Pagan rulers.[85] Lesser known is the fact that while the court of Burma was undergoing a religious revival under the auspices of the 'Chindwin scholars', a group of Buddhist priests and lay literati under the Konbaung rulers, it was Brahmins from Manipur who contributed to '... shaping the perspectives of and cooperating in the literary activities of Chindwin-based Buddhist scholars and laypeople'.[86] This example also shows that while the migration of people into Northeast India was from and through Burma, reverse flows also occurred, especially in the realm of religious exchange.

Such adoptions were no doubt the result of the circulation of ideas and religious personnel who propagated these ideas. Vaishnavism in Manipur had been the result of the arrival of Brahmins (Goswamis) from Bengal into Manipur, travelling west to east.[87] Moving in the opposite direction, Shankardeva, who initiated the neo-Vaishnava movement in Assam in the sixteenth century, had been on pilgrimages to places as far as Orissa, besides north India, thus broadening the religious networks and the sacred geography that the people of the Northeast had been a part of. Islam, too, had made its presence felt in both Assam (thirteenth century) and Manipur (seventeenth century).

Even outside of these valley states, the hills were not beyond the reach of these religious networks. Buddhism has had a presence in the hills and appears to have arrived from Tibet and Burma.[88] Although many of the communities in western Arunachal Pradesh such as the Monpas and the

83 Guha, 'Ahom', p. 166.
84 Harvey, *History*.
85 Pearn, *Burma*.
86 Charney, 'Literary Culture', p. 162.
87 Sinha, *Tribal Polities*, p. 145.
88 Singh, 'North-east India', pp. 257-282.

Sherdukpens believe that Buddhist emissaries from Tibet arrived in their region as early as during the eighth century at around the time when Tibet itself was being introduced to Buddhism, the Tawang monastery which represented an institutionalized presence was established in the seventeenth century. Another group of Buddhist followers in eastern Arunachal Pradesh (Khamptis, Singphos, Tangsas) brought it with them as they migrated into the region in the eighteenth century. The former were followers of the Mahayana or Lama traditions of Buddhism while the latter were Theravada Buddhists. At another location, in the expedition to the Lushai Hills in the late nineteenth century, Woodthorpe remarked that a Burmese idol had been found in the tomb of Vonolel, one of the great chiefs of the Lushai Hills whose village stood on the Lushai-Chin border.[89] Similarly, the British Expedition of 1889-90 noted the presence of a Buddhist monk in the Chin Hills who had come to look for his brother who had been captured.[90] N. Chatterji has also pointed to three rock-cut engravings in Mizoram in Mualcheng, some 65 kilometres south of Lunglei. The meditational pose of one of them suggests that they are either Hindu or Buddhist, and Chatterji deems the latter more probable.[91] Unfortunately, these images have not been dated, and they differ from those normally found in Mizoram, although similar engravings have been found in a few other places in Mizoram. The location of these engravings suggests that Mualcheng, being relatively close to the Chittagong Hill Tracts (which were predominantly Buddhist), came within the latter's radius of influence.

Another image from Suangpuilawn in Mizoram, close to the Manipur border, represents a deity on a horse/pony/mule, which combines the earlier idea of circulation of equines with the idea of religious mobility. Many of these instances do not necessarily mean the adoption of the religion by the indigenes, but they do point to the circulation and movements of these beliefs through itinerants.

In the late nineteenth and twentieth centuries, even as Christianity became an important aspect of the religious landscape of the region, Christianity was not largely spread through western missionaries but instead through locals who felt conviction about their faith and the need to propagate it. For example, much of the Christianity amongst the so-called Old-Kuki groups was introduced by missionaries from the Lushai Hills. In

89 Woodthorpe, *Lushai Expedition*. It was remarked that he had come into the Hills without an escort, and he was even able to ransom back his younger brother from the hill tribes, p. 282.

90 Reid, *Chin Lushai*, p. 174.

91 Chatterji, *Monoliths*, pp. 35-36.

Figure 5.3 Rock engravings at Mualcheng, Mizoram

turn, many Khasis were also part of the propagation of Christianity to the Lushai Hills.[92]

The importance of the second regime, courtly exchanges, is evident throughout the Northeast. According to Michael Charney, the exchanges between the royal courts of Manipur and Burma had increased from the mid-seventeenth century with the latter's expansion and incursions into the Upper Chindwin or Kabaw valleys. Since the time of the Konbaung ruler Alaunghpaya (r.1752-60) and following such wars, there have been examples of cultural flows from Manipur into the Chindwin Valley. It was said that Alaunghpaya 'deported thousands of Manipuri boatmen, silversmiths, and silk weavers to the royal capital area,'[93] who themselves were influenced by Bengal. Similarly, it was also said that Choorjeet Singh, one of the princes of Manipur who had fled to Ava in 1806, and his principal officers 'had from their long residence in Ava adopted many of the luxurious habits of that court, and they affected a degree of splendour in dress and equipment of their horses, which contrasted very strikingly with their former simplicity.'[94]

What these royal connections and internecine warfare suggest is the existence of a network of routes that passed through deep jungles and

92 Pachuau, 'Assistants'.
93 Charney, 'Literary Culture', p. 169
94 Pemberton, *Report*, p. 45.

mountainous terrain. For instance, there were said to be three land routes between Arakan and Chittagong.[95] Pemberton also talks about three different routes that connected Cachar and Sylhet with Manipur and the latter with Burma.[96] Pemberton additionally mentioned the many routes from Manipur to Ava, some of them passing through the territories of 'Tangkool tribes' as well as the 'Anal and Mueeyol tribes of Nagas.'[97] It was also probably through the Patkai pass that the Shans and the various Singpho tribes first conquered Assam.[98]

All this is evidence that the states that existed in the pre-colonial period, at various points in time, disturbed the political equilibrium and movements of people – whether royal or otherwise – was much more common than current historiography alludes to or acknowledges.

Another means of looking at courtly connections and exchanges across the region is through the perspective of marital alliances. The valley states were not merely sites for political and religious exchanges. Marital alliances were an important aspect of statecraft amongst the so-called established states. There are many examples of bride exchanges amongst the royal households. Pemberton reported that in 1798 'Jaee Sing' of Manipur had given his daughter in marriage to the Tripura Raja.[99] Pemberton also alludes to a marriage alliance as early as in the fifteenth century between the Pong (one of the Shan chiefs of Burma) rulers and Manipur who together attacked the Burmese rulers at Ava.[100] Upon the arrival of the British in the Jaintia hills, Pemberton also mentions that the Jaintia rulers, who had stronger relations with the people on the plains, were allowed to marry women of two particular castes from the plains, namely those of the 'Kayt' and 'Bayd' castes.[101] As is seen in the context of South Asia, the creation of legitimacy through marital alliances, formalization of religious practices, and adoption of cultural practices associated with more established royal lineages – all important aspects of statecraft and state formation – also took place across the states in the region under survey.

Despite the emphasis that has existed in historiography regarding the raids by hill men on the plains, much trade took place between the two

95 Pemberton, *Report*, p. 95
96 Pemberton, *Report*, p. 51
97 Pemberton, *Report*, p. 56.
98 Pemberton, *Report*, p.64.
99 Pemberton, *Report*, p. 43.
100 Pemberton, *Report*, p. 112.
101 Pemberton, *Report*, p. 219. 'Kayt' and 'Bayd' castes here refer to the kayastha and baidya castes, who along with the brahmanas constituted the upper caste cluster of Bengal.

zones, especially through rivers, and this is the third regime that I am analysing. The dealings between Bengal and Assam were mainly through the Brahmaputra, and boats of various sizes could move easily between the Ganges and the Brahmaputra.[102] Assam's exports included lac, cotton, mustard seed, silk, gold bullion, rice, wax, black pepper, long pepper, Indian madder, *ao tenga* (*dillenia speciosa*), betel nuts, and rattan, which were largely products brought to Assam by the hill communities. Assam's main import was salt, besides fine muslins and calicoes from Bengal. European products such as English woollens, taffetas, and European cutlery also came to Assam through Bengal.[103] The Barak, the river which had as its source the hill country surrounding the Manipur valley, flowed out onto the plains in Sylhet where three bazaars were located: Buxie-Bazar, Hubbegunjee, and Nubbegunjee. These were the principal emporia and according to Pemberton, 'to these points, boats of 1,000 maunds burthen annually repair, laden with the produce of the more western districts, and principally carry off in exchange the surplus rice of Cachar and Sylhet, and the timber, bamboos, and grass, which are abundantly produced in the forests on their eastern and southern borders.'[104] Cachar was an important source for various types of timber, of which *jarul*[105] was used for boat building in Sylhet. It was brought down the Barak and Jiri rivers. Because of its buoyancy, *jarul* was well suited to make large chunam boats,[106] which could last for ten to twelve years.[107] At the Lushai bazaars, Mackenzie noted that rice, red cloth, and brass vessels were exchanged.[108] Similarly, on the 'Currumfullee' (Karnafuli), Pemberton wrote that the 'Bengalees are in the habit of penetrating, during the cold weather, four or five days' journey, by water, beyond the Burkal falls for the purpose of cutting bamboos and killing wild elephants.'[109] Woodthorpe also remarked that a 'Cabuli' (from Kabul) fruit seller was found in Chepui in the Lushai Hills at the time of the early expeditions into the Lushai Hills (late nineteenth century).[110] Similarly, Mackenzie also wrote about the presence of 'Meerware' (Marwari) merchants from western India who imported broad cloths, muslin, salt, opium, liquor, glass,

102 Robinson, *Descriptive Account.*

103 Robinson, *Descriptive Account*, pp. 238-243.

104 Pemberton, *Report*, p. 4.

105 *Lagerstroemia speciosa*, also known as Pride of India, Queen's crepe-myrtle.

106 Boats in which the caulking material is made of *chunam* or lime.

107 Fisher, 'Timber Trade', pp. 205-206.

108 Mackenzie, *Northeast Frontier*, p. 418.

109 Pemberton, *Report*, p. 6

110 Woodthorpe, *Lushai Expedition*, p. 310

crockery, tobacco, betel, and rice for gold dust, gold, ivory, silver, amber musk, *daos* (Burmese cutlasses were found in the interiors of the Arakan *yoma* or mountains[111]), Burmese cloths, and small Chinese boxes.[112] It was also noted by Mackenzie that cotton had been exchanged from the hills since 1833, while 'Chinamen' were the main traders on the Irrawaddy rather than the 'indolent' Burmese themselves.[113] Striped silk cloths, said to be the 'delight' of the Khasis, were imported from the Assam plains as well as the neighbouring Jaintia Hills where they were manufactured from silk made from the eri silkworm reared by the Assamese and the Jaintias.[114] Bhutan was also a great market for the two varieties of Assamese silk, *eri* and *muga*.[115]

Tobacco, cotton, ginger, and pepper were cultivated by all the hill people according to Pemberton, and the tribes bordering on the plains of Assam, Bengal, and Ava conducted limited trading of these products with the inhabitants of these countries. It was said that the Lushais had always been trading in the Chittagong marts.[116] On the other hand, those who occupied the central ranges (of the Arakan arc) and who had no direct dealings with the inhabitants of the plains bartered the produce of their hills with the next adjoining tribe. Thus, according to Edgar, salt was traded by the Lushais from the western plains to the Howlongs in the eastern reaches of the Lushai Hills district in exchange for rubber, with the former being four times its weight in rubber.[117] Bell-metal gongs and 'kurtals' (cymbals), manufactured by the inhabitants of Yunnan, were also found in almost all the hill villages along the eastern frontier, according to Pemberton, suggesting the extent of exchanges across the region. Similarly, according to Hall, Burma was trading with Yunnan for cotton. Iron ore was also found at intervals throughout the hills.[118] In the Khasi Hills specifically, however, according to Yule, the evidence of it being smelted was so extensive that the smelting activities could have taken place for at least twenty centuries.[119] The iron was then 'exported from the hills in the shape of hoes to the Assam valley and in lumps of pig iron to the Surma valley, where it was used by boat builders

111 British Library L/PS/6/547, p. 33
112 Mackenzie, *Northeast Frontier*, pp. 71-73.
113 Reid, *Chin Lushai*, p. 82.
114 Gurdon, *Khasis*, p. 59.
115 Robinson, *Descriptive Account*, p. 243
116 Mackenzie, *Northeast Frontier*, p. 433.
117 Mackenzie, *Northeast Frontier*, p. 433
118 Hall, *History*, p. 619.
119 Yule, 'Notes', p. 853.

for clamps.'[120] In 1853 it was estimated that 20,000 maunds of iron were exported from the Khasi Hills. The practice, however, was dying out by the early twentieth century to such an extent that the forges had to use rough iron brought from the plains.[121]

This study highlights the kinds of networks that have existed in a region that, in historiography, has generally been considered isolated and inaccessible. The evidence above, by no means exhaustive, suggests instead the movement of people, goods, and ideas in and out of the region at different points of time in history in the pre-colonial past. More importantly, I have also suggested that these connections were not always in the immediate vicinity but that the region has been included in long-distance networks that have shaped the region. For instance, the widespread use of firearms, even though they were not locally manufactured, and the evidence of movement of people and goods also suggest the region possessed vitality and dynamism. At the time of the arrival of the British, we find some well-established states, which acted as magnets that engendered these connections. At the same time, we find tribal groups warring with each other and vying for supremacy over others, groups that were attempting to carve out and thereby create their own stable systems of governance. The ability to procure resources, especially firearms, went a long way in establishing this ascendancy, something that the British corroborated. The procurement of cowries also facilitated the establishment of social cleavages within a given society even as they became markers of wealth and honour.

Furthermore, it is interesting to note the far-reaching connections that people in the region were part of – from relations with neighbouring Yunnan, Burma, and Bengal (horses and cowries), to further afield, albeit indirectly, Europe, the Americas, and the Philippines (firearms and cowries). The region was also part of a religious landscape that brought about connections with the Gangetic plains of India as well as Tibet and across the great Indo-Burman arc into the Irrawaddy valley and back. The incorporation of more formalized religions not only ensured legitimacy of the rulers but furthered social formation through the creation of divisions in society. The resources that these webs of connections engendered, whether ideological or material, definitely formed the region in the pre-colonial period, the scale of which is something rarely acknowledged.

120 Gurdon, *The Khasis.* p. 57

121 Gurdon, *The Khasis.* p. 58

References

Ahmed, Imdad Ali. 'Elephants in Medieval Assam'. *Proceedings of the Indian History Congress* 76 (2015): pp. 264–270.

Allan, J. 'The coinage of Assam'. *The Numismatic Chronicle and Journal of the Royal Numismatic Society* (1909): pp. 300–331.

Bareh, Hamlet. *The History and Culture of the Khasi People*. Guwahati: Spectrum Publications, 1997.

Basic Statistics of North Eastern Region 2015, Government of India, North Eastern Council Secretariat, Shillong, accessed 2 December 2021, http://necouncil.gov.in/sites/default/files/uploadfiles/BasicStatistic2015-min.pdf

Bhattacharya, Neeladri, and Joy L. K. Pachuau, eds. *Landscape, Culture and Belonging: Writing the History of Northeast India*. Cambridge: Cambridge University Press, 2019.

Blackburn, Stuart. 'Memories of Migration: Notes on Legends and Beads in Arunachal Pradesh, India'. *European Bulletin of Himalayan Research* 25, no. 26 (2003): pp. 15–60.

Blackburn, Stuart. *Himalayan Tribal Tales: Oral Tradition and Culture in the Apatani Valley*. Leiden: Brill, 2008.

Carey, Bertram Sausmarez, and Henry Newman Tuck. *The Chin Hills: A History of the People, our Dealings with them, their Customs and Manners, and a Gazetteer of their Country*, vol. 1. Rangoon: Superintendent, Government Printing, Burma, 1896.

Chang, Wen-Chin. *Beyond Borders: Stories of Yunnanese Chinese Migrants of Burma*. Ithaca, NY and London: Cornell University Press, 2014.

Charney, Michael W. 'Literary Culture on the Burma–Manipur Frontier in the Eighteenth and Nineteenth Centuries'. *The Medieval History Journal* 14, no. 2 (2011): pp. 159–181.

Chatterji, N. *Monoliths and Landmarks of Mizoram*. Calcutta and Aizawl: Firma KLM and TRI, 1979.

Chaube, Shibani Kinkar. *Hill Politics in Northeast India*. New Delhi: Orient Longman, 1999.

Chew, Emrys. *Arming the Periphery: The Arms Trade in the Indian Ocean During the Age of Global Empire*. Basingstoke: Palgrave Macmillan, 2012.

Dzüvichü, Lipokmar. 'The Colonial State and "Illegal" Arms Trade along the North East Frontier of British India, 1860s–1900s'. In *Landscape, Culture and Belonging: Writing the History of Northeast India*, edited by N. Bhattacharya and J. L. K. Pachuau. Cambridge: Cambridge University Press, 2019.

Endle, Sidney. *The Kacharis*. London: Macmillan, 1911.

Fisher, Thomas. 'Timber Trade in Cachar'. *Journal of the Asiatic Society of Bengal* 1 (1832): pp. 305–306.

Goodenough, F. A. 'Letter to Major-General Sir Andrew Scott Waugh, on Routes between Upper Assam and Western China'. *Proceedings of the Royal Geographical Society of London* 12, no. 5 (1867–68): pp. 334–336.

Guha, Amalendu. 'The Ahom Political System: An Enquiry into State Formation in Medieval Assam, 1228-1800'. In *Tribal Polities and State Systems in Pre-Colonial East and Northeastern India*, edited by Surajit Sinha. Calcutta: KP Bagchi and Co., 1987.

Gurdon, Philip Richard Thornhagh (Major). *The Khasis*. London: David Nutt, 1907.

Hall, Daniel George Edward. *A History of South-East Asia*. London and New York: Macmillan and St. Martin's Press, 1968.

Harvey, G. E. *History of Burma*. Calcutta: AES, 2000.

Hazarika, Nilam. 'A Study into the Ahom Military System in Medieval Assam'. *International Journal of Humanities and Social Science Invention ISSN* 4, no. 6 (2015): pp. 17–22.

Henniker, F. C. *Gold and Silver Wares of Assam: A Monograph*. Shillong: Assam Secretariat Printing Office, 1905.

Horam, M. *Social and Cultural Life of Nagas (The Tangkhul Nagas)*. New Delhi: BR Publishing Corporation, 1977.

Hussain, Syed Ejaz. 'Silver Flow and Horse Supply to Sultanate Bengal with Special Reference to Trans-Himalayan Trade (13th-16th Centuries)'. *Journal of the Economic and Social History of the Orient* 56, no. 2 (2013): pp. 264–308.

Hutton, John Henry. *The Angami Nagas*. London: Macmillan, 1921.

Hutton, John Henry. 'Naga Chank Ornaments of South Indian Affinities'. *Man* 26 (1926): pp. 222–224.

Jhala, Angma Dey. *An Endangered History: Indigeneity, Religion, and Politics on the Borders of India, Burma, and Bangladesh*. New Delhi: Oxford University Press, 2019.

Kanungo, Alok Kumar. 'Naga Ornaments and the Indian Ocean'. *Bulletin of the Indo-Pacific Prehistory Association* 26 (2006): pp. 154–162.

Laldinpuii, K. 'A Memorial Stone Platform of Lapuithanga Sailo Chief'. In *A Modern History of Mizoram*, edited by Sangkima. Guwahati: Spectrum Publications, 2004.

Lehman, F. K. *The Structure of Chin Society: A Tribal People of Burma Adapted to a Non-Western Civilization*. Urbana, IL: University of Illinois Press, 1963.

Lewin, Thomas Herbert. *The Hill Tracts of Chittagong and the Dwellers Therein; With Comparative Vocabularies of the Hill Dialects*. Calcutta: Bengal Printing Company, 1869.

Lotha, Abraham. 'Articulating Naga Nationalism'. PhD diss., City University of New York, 2009.

Ludden, David. 'The First Boundary of Bangladesh on Sylhet's Northern Frontiers'. *Journal of the Asiatic Society of Bangladesh* 48, no. 1 (June 2003): pp. 1–54.

Ma, Jianxiong., and Cunzhao Ma. 'The Mule Caravans of Western Yunnan: An Oral History of the Muleteers of Zhaozhou'. *Transfers* 4, no. 3 (2014): pp. 24–42.

Mackenzie, Alexander. *The Northeast Frontier of India,* [first published as *History of the Relations of the Government with the Hill Tribes of the Northeast Frontier of India,* 1884]. Delhi: Mittal, 1979.

Malsawmliana. 'Megalithic Monuments of the Mizos'. *Historical Journal Mizoram* VI (2005): pp. 52–62.

Markovits, Claude, Jacques Pouchepadass, and Sanjay Subrahmanyam. *Society and Circulation: Mobile People and Itinerant Cultures in South Asia, 1750-1950.* London: Anthem Press, 2006.

M'Cosh, John. *Topography of Assam.* Calcutta: Bengal Military Orphan Press, 1837.

McCosh, John. 'Account of the Mountain Tribes on the Extreme N.E. Frontier of Bengal'. *Journal of the Asiatic Society of Bengal* 203 (1836).

Mills, John P. 'The Mishmis of the Lohit Valley, Assam'. *The Journal of the Royal Anthropological Institute of Great Britain and Ireland* 82, no. 1 (1952): pp. 1–12.

Pachuau, Joy L. K. *Being Mizo: Identity and Belonging in Northeast India.* New Delhi: Oxford University Press, 2014.

Pachuau, Joy L. K., and Willem van Schendel. 'Borderland Histories, Northeastern India: An Introduction'. *Studies in History* 32, no. 1 (2016): pp. 1–4.

Pachuau, Lalsangkima. '"Assistants" or "Leaders"? The Contributions of Early Native Christian Converts in Northeast India'. In *Christianity in Indian History,* edited by P. Malekandathil, J. L. K. Pachuau, and T. Sarkar. Delhi: Primus, 2015.

Parratt, Saroj N. Arambam. *The Court Chronicle of the Kings of Manipur: The Cheitharon Kumpapa, Original Text, Translation and Notes, vol. 1, 33-1763 CE.* London and New York, NY: Routledge, 2005.

Parry, Nevill Edward. *The Lakhers.* London: Macmillan and Co. Ltd., 1932.

Pearn, Bertie Reginald. *Burma Background* (No. 1). London: Longmans Green & Co., 1943.

Pemberton, Robert Boileau. *Report on the Eastern Frontier of British India.* Calcutta: Baptist Mission Press, 1835.

Pommaret, Françoise. 'Ancient Trade Partners: Bhutan, Cooch Bihar and Assam' (17th to 19th centuries)' (2000), accessed 2 December 2021, http://himalaya.socanth.cam.ac.uk/collections/journals/jbs/pdf/JBS_02_01_02.pdf

pp. 1–25.

Rajesh, M. N. 'Tibet and Bengal, Trade Routes and Long Distance Exchange with Reference to Silver'. In *From Mountain Fastness To Coastal Kingdoms,* edited by J. Deyell and R. Mukherjee. Delhi: Manohar, 2019.

Reid, Adam Scott. *Chin-Lushai Land.* Calcutta: Thacker, Spink and Company, 1893.

Robinson, William. *A Descriptive Account of Assam: With a Sketch of the Local Geography and a Concise History of the Tea-plant of Assam*. London: Ostell and Lepag, 1841.

Rosanga, O. 'The Consolidation of Mizoram with Special Reference to the Disarmament of the Hills'. *Historical Journal Mizoram* VIII (2007 [1929]): pp. 98–108.

Shakespear, John. *The Lushei Kuki Clans*. London: Macmillan, 1912.

Shakespear, Leslie Waterfield (Colonel). *History of the Assam Rifles*. London: Macmillan, 1929.

Shimray, R. R. *Origin and Culture of Nagas*. New Delhi: Somsok Publications, 1985.

Singh, B. P. 'North-East India: Demography, Culture and Identity Crisis'. *Modern Asian Studies* 21, no. 2 (1987): pp. 257–282.

Sinha, Surajit, ed. *Tribal Polities and State Systems in Pre-Colonial East and North-eastern India*. Calcutta: KP Bagchi and Co., 1987.

Stevenson, Henry Noel Cochran. *The Hill Peoples of Burma* (No. 6). London: Longmans, Green & co., Ltd., 1945.

Stonor, C. R., and Edgar. Anderson. 'Maize among the Hill Peoples of Assam'. *Annals of the Missouri Botanical Garden* 36, no. 3 (1949): pp. 355–404.

Thohe Pou, R. B. 'The Myths of Naga Origin', published 2007, accessed 2 December 2021, https://thohepou.wordpress.com/2007/08/17/the-myths-of-naga-origin/.

Van Schendel, Willem. 'Geographies of Knowing, Geographies of Ignorance: Jumping Scale in Southeast Asia'. *Environment and Planning D: Society and Space* 20 (2002): pp. 647–668.

Vogel, Hans Ulrich, and Sabine Hieronymus. 'Cowry Trade and Its Role in the Economy of Yunnan: From the Ninth to the Mid-Seventeenth Century'. *Journal of the Economic and Social History of the Orient* 36, no. 3 (1993): pp. 211–252.

Woodthorpe, Robert Gosset. *The Lushai Expedition, 1871-1872*. London: Hurst and Blackett, 1873.

Yang, Bin. 'Horses, Silver, and Cowries: Yunnan in Global Perspective'. *Journal of World History* 15, no. 3 (September 2004): pp. 281–332.

Yang, Bin. 'The Bay of Bengal Connections to Yunnan'. In *Pelagic Passageways: The Northern Bay of Bengal Before Colonialism*, edited by Rila Mukherjee, pp. 317–342. Delhi: Primus Books, 2011.

Yang, Bin. 'The Rise and Fall of Cowrie Shells: The Asian Story'. *Journal of World History* (2011): pp. 1-26.

Yang, Bin. 'The Laws of King Mengrai, Cowrie Shells and Intra-Asian Interactions'. In *From Mountain Fastness to Coastal Kingdoms*, edited by J. Deyell and R. Mukherjee. Delhi: Manohar, 2019.

Yule, Henry. 'Notes on the Iron of the Kasia Hills, for the Museum of Economic Geology'. *Journal of the Asiatic Society of Bengal* 11, no. 2 (1842): pp. 853–857.

About the author

JOY L. K. PACHUAU, Professor of History, Centre for Historical Studies, Jawaharlal Nehru University, New Delhi, India. Her research includes the socio-cultural history of Northeast India, and the history of religion and society in Portuguese expansion in Asia. Among her publications are *Being Mizo* (2014) and *The Camera as Witness* (2015, with van Schendel).

6 Flows and Fairs

The Eastern Himalayas and the British Empire

Arupjyoti Saikia

Abstract

The Eastern Himalayas and the Patkai mountain range, which form the northern frontier areas of Assam, have hosted numerous pathways connecting Assam with Burma and Southwest China. For centuries, they sustained a complex network of commercial exchange across diverse polities and economies. Local economic, environmental, and political conditions influenced these seasonal flows. Things began to change in the nineteenth century, and the colonial government sustained the commercial networks by political, legal, and economic means until the battles of the Second World War reached Assam. This chapter discusses how the colonial government, despite its powerful bureaucratic structures and military might, resorted to multiple methods to keep these inherited flows of commerce in motion.

Keywords: Assam, Himalayan foothills, fairs, rivers, plants

The Eastern Himalayas and the Patkai, which form the northern frontier areas of Assam, have hosted numerous pathways across the geographical landscape in Southwest China, Burma, and Assam. For centuries, they sustained a complex network of commercial exchange across diverse polities and economies. Local economic, environmental, and political conditions influenced these seasonal flows. In the nineteenth century, they opened up new opportunities for English and Indian commerce, and the merchants lost no opportunities to engage after the British East India Company (EIC) annexed Assam in the 1820s. The aim to maintain control and government for securing British trade was challenged by people resident in the mountain areas who confronted the commercial operations. Indian merchants, in

Cederlöf, Gunnel, and Willem van Schendel (eds), *Flows and Frictions in Trans-Himalayan Spaces: Histories of Networking and Border Crossing*. Taylor & Francis Group 2022
DOI: 10.5117/9789463724371_CH06

Map 6.1 Market towns connecting central and north Assam, eastern Tibet, Sichuan in China and Kachin, 19th and early 20th century

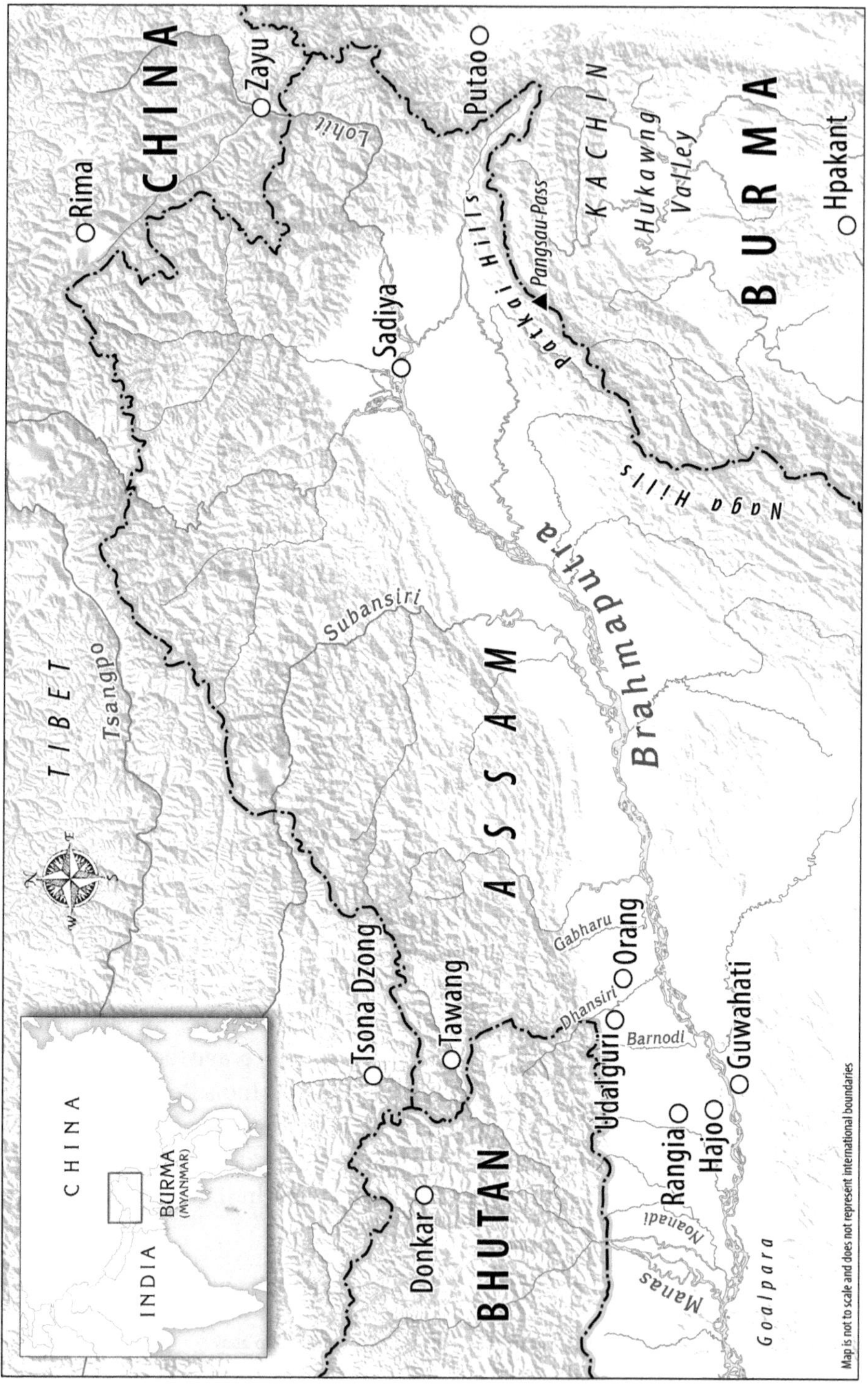

Drawn by Laurie Whiddon

contrast to the English, successfully established a complex network of social and economic relationships with the mountain residents. This facilitated the growth of the Indian merchant capital. However, the colonial government sustained the commercial networks through political, legal, and economic means until fighting in the Second World War reached Assam, and the frontier areas became central scenes of warfare. After the war, commercial activities declined rapidly.

This chapter discusses how the colonial government, despite its overwhelming bureaucratic structures and military might, resorted to multiple methods to keep the flows of commerce that it inherited in motion. The study suggests that throughout the nineteenth century, while remaking Bengal's eastern frontiers and Assam's economy and polity, the colonial government continuously reproduced, sustained, and fortified these commercial flows. While the colonial government invested in and exploited the lowlands of Assam, it continued to utilize the commercial flows by exploring the trade routes, hunting for natural and artisanal commodities, and striving to govern and bring these seemingly unwieldy economic activities into the orbit of the British Empire. They thereby re-energized the economic activities of the mountain and lowland areas. However, the colonial government only became a partial custodian of the expansive geography. Under the British government, activities were mainly restricted to the eastern Himalayan foothills, whereas in the uplands, human mobility and commercial transactions mostly retained their earlier forms while also mixing with commercial activities within the British Empire. By the end of the nineteenth century, economic activities in the lowlands, south of the foothills, were firmly integrated into the Bengal economy through modern transport infrastructure. Because of this, the commercial value of mountain products increased as they were in high demand on a larger market. As a result, the characteristics of commercial networks changed and gave way to new ones. Colonial intervention thereby ensured that the old routes and new networks complemented each other and merged, which eventually re-energized the old commercial networks between the eastern Himalayas and Assam via Bengal and the global commercial actors, about which much has already been written.[1] However, we are yet to understand how Assam connected to the eastern Himalayas. This study therefore examines the trade networks that the EIC eventually took control of in the early nineteenth century.

1 Ludden, 'Process', pp. 132-150; Ludden, 'Cowry Country', pp. 75-100; Malekandathil, 'Circulation', pp. 259-286; Cederlöf, *Founding an Empire*; Cederlöf, 'Seeking', pp. 125-144; Yang, 'Horses', pp. 281-332.

Commercial networks

> ... Assam formed a highway not only for trade but also for the exchange of ideas between India and Burma and Southwest China ... from at least the closing centuries of the first millennium B.C.[2]

In 1954, the Indian linguist Suniti Kumar Chatterji wrote about the importance of Assam as 'a highway between India and China' through 'pestilential jungles and high mountains', and sinologists concurred.[3] When the EIC conquered Assam, they were aware of routes that 'linked areas across these huge societies, and some connected to places still father away'.[4] Such routes were superimposed onto existing natural features including the river channels.[5] During monsoons, routes were never permanent, while in winters, when the rivers had 'dry stony beds', routes opened up everywhere, and the mountain residents used them to reach the plains.[6] Products from biodiverse environments were brought by traders through the routes that crossed diverse political, cultural, religious, and ecological milieus. Travelling downhill, they added local merchandise. For some travellers, the journey was a pilgrimage. Hajo on the river Brahmaputra was one of those sites. Surrounded by vibrant textile-producing artisan-peasant settlements, it was a crucial junction where trade and pilgrimage intermingled.[7] 'The pious Budhist [sic] too, imbued with the same faith, leaves his home in the distant regions of China and Thibet ... hastens to make obeisance at the shrine of his country's deity', wrote the British officer William Robinson in 1841.[8]

While following the rhythms of life, the mountain routes and lowland societies were strictly regulated.[9] The foothills formed a 'multistate margin' where highlanders and lowlanders interacted and gave shape to a 'multistate space' where people were restricted by their respective yet fluctuating

2 Chatterji, *Place*, p. 16.

3 Yang, 'Horses', p. 283.

4 Sumit and Pomeranz, 'Life', p. 59.

5 Cederlöf, *Founding an Empire*.

6 Williamson, 'Lohit-Brahmaputra', p. 365; Assam, *Report on the Administration of the Province of Assam, 1875-76*, p. 8.

7 On Tibetan pilgrimage to Hajo, see Waddell, *Buddhism*, pp. 308-314; Ray, *Trade*, pp. 91-98. On the economic activities and peasant-artisanal productions in the larger region, see Guha, 'Medieval Economy', pp. 478-505.

8 Robinson, *Descriptive Account*, pp. 259.

9 Christian, 'Trans-Burma Trade Routes', pp. 173-191; Giersch, 'Cotton'; van Schendel, *Bengal Borderland*; Cederlöf, 'Corridors'.

jurisdictions.[10] Using a shared mechanism, the rulers of Bhutan, Tibet, and Assam kept the mercantile exchanges under checks and balances. Through different political systems, each strove to enforce their own regulations in the eastern Himalayan foothills. Economic exchanges between Bengal, Assam, and Bhutan merchants were regulated at the customs checkpoints (*duars*). Francis Hamilton, the EIC surgeon, surveyor, and botanist, estimated that there were 18 such *duars* in the 1910s.[11] The control of forms for sharing and deciding on the collection of taxes at the *duars* oscillated between Assam and Bhutan. The Assam rulers controlled the *duars* during the monsoon when they reclaimed the areas for agricultural activities. Frontier officials imposed strict vigilance and ensured that merchants abided by the political agreements. An appointed broker regulated purchase and sale. The diverse currency environments in the larger region meant that merchants had wider choices and advantages.[12] Duties collected at markets formed part of the imperial finance of Assam.[13] During the rest of the year, these tracts were under the Bhutan government that paid annual tributes to Assam. The tributes acted as a medium of exchange and a mechanism for sustaining a trade network between highlands and lowlands. They included highly prized mountain products such as yak tails, ponies, musk, gold dust, blankets, and knives, gathered from within the Himalayas. Merchants from Tibet and Bhutan brought rock salt, woollens, gold dust, horses, *chowries*, and Chinese silk.[14] Tibetan trade was mainly in the hands of Kumpas, the itinerant Tibetan traders.[15] Lowland merchants brought goods collected from the floodplain economies and distant Indian markets via the Dooars. The products included rice, dried fish, Muga silk, coarse silk cloth woven by Assamese women, iron and lac, buffalo horns, pearls, and corals .[16] Some trade flows went in other directions, as did the elephants procured by the Mughal army. They were transferred either as part of diplomatic treaties with the local rulers or bought from merchants.

In Assam, the flows followed an intricate river transport network, either via the artery of the river Brahmaputra to Bengal or uphill and across

10 Smyer Yü, 'Introduction', pp. 11-42; Shneiderman, 'Central Himalayas', pp. 289-312.

11 Hamilton, *Account*, p. 68; Mackenzie, *History*, pp. 9-10.

12 Ludden, 'Cowry Country', p. 78; Pommaret, *Ancient Trade Partners.*

13 Gait, *History*, p. 207.

14 Devi, *Ahom-Tribal Relations*, p. 250.

15 Robinson, *Descriptive Account*, p. 347; Hunter, *Statistical Account*, p. 143-145.

16 Devi, *Ahom-Tribal Relations*, p. 250; Robinson, *Descriptive Account*, p. 347; M'Cosh, *Topography*, p. 10; Hamilton, *Account*, p. 74.

Meghalaya to Sylhet on the Surma in eastern Bengal.[17] In the early nineteenth century, these extensive routes began to collapse to merge with the larger trade systems operated by the EIC. As a result, the number of custom checkpoints and passes decreased towards the east.[18] Hamilton estimated the value of trade to 200,000 rupees in the 1810s. Given the financial crisis at that time, it was still a significant amount.[19] The trade network was a site of intense rivalries between traders, which necessitated state regulation and diplomatic negotiation. A diplomatic standstill agreement [*posa*] between the highland and lowland residents, which increased over two centuries, allowed people in the highlands to extract an annual tribute from people in the lowlands.[20] The agreements worked to stabilize the political flux and sustained cross-border economic activities.

Cross-cultural flows facilitated mobility of ideas and cultural forms which became part of the commercial networks. Tibetan interest in Assamese textiles is illustrative. The *Vrindavani Vastra*, a 'figured silk textile' produced under the direction of Śankaradeva (died in 1568 CE), the Assamese Vaishnavite saint scholar, was part of this culture of exchange.[21] Several fragments of the Vrindavani Vastra were later recovered in Tibetan Buddhist painting to become a revered cultural object in Tibet. There were also instances of Assamese interest in the Tibetan material and cultural world.[22] There are depictions of scenes of diseases on the folio of the Assamese manuscript *Brahmavaivarta Purāṇa* from the early nineteenth century that were 'remarkably close to similar illustrations in Tibetan medical manuscripts', according to Phyllis Granoff.[23]

Linguists have observed a complex interplay of linguistic crosscurrents into Assamese, one of the main languages spoken in the south of the foothills.[24] In 1872, an English colonial official amateurishly noted how the Assamese language had a 'mixture of Indo-Chinese blood'.[25] Such views were rigorously re-examined later. Banikanta Kakati, the Indian linguist who relied heavily on the Indo-Aryan linguistic roots to understand the rise and growth of the Assamese language, accepted that the languages

17 Ludden, 'Cowry Country'; Saikia, *Unquiet River*, Ch. 3; Guha, 'Medieval Economy', p. 490.

18 Ahmad, 'Assam-Bengal Trade', p. 176; Pemberton, *Report*, p. 78.

19 Hamilton, *Account*, p. 73-74; Richards, 'Finances', Table 5.

20 Devi, *Ahom-Tribal Relations*, p. 199-200; Guite, *Against State*, p. 250-263.

21 Blurton, *Krishna*.

22 Granoff, 'Illuminating the Formless'.

23 Ibid., 131.

24 Grierson, *Linguistic Survey*, p. 1.

25 Waterfield, *Memorandum*, p. 20.

in the Eastern Himalayas had a major influence on the rise of Assamese.[26] Kakati's studies, and linguists who followed him in studying archives of the Tibetan language, reconfirm the influence from continuous human mobility.

Search for commercial routes

When the EIC took control of Assam, their eyes were on these routes. As early as in the eighteenth century, European merchants in India strove to enter China via Tibet in a broader search for natural products. However, diplomatic and commercial results were mixed. Henry Thomas Colebrooke, the English orientalist, remarked that 'many dyes and medicinal drugs ... now imported into England' could be supplied from these regions.[27] In the 1750s, Jean-Baptiste Chevalier, an officer in the French East India Company, unsuccessfully parleyed in the Tibetan court to seek a trade mission to China.[28] Towards the end of the eighteenth century, English traders followed in the tracks of Indian merchants to find the sources of trade.[29] The Scottish merchant Robert Lindsay, who earned substantial profits from limestone and conversion of cowries as a trader and as Collector of Sylhet, noted how the mountains south of the eastern Himalayan foothills provided 'elephants of the best description'.[30]

After the conquest of Assam and victory over Burma in the 1820s, maps were drawn that depicted markets, lists of merchandise, and commercial and military routes for entering the commercial world in the east. Among those was a map published in 1825 by an officer named Wood with detailed descriptions of routes from Assam to Bhutan.[31] Lieutenant John Bryan Neufville prepared a list of twenty trading points between Assam and Tibet.[32] However, the EIC met with fierce defiance from the mountain residents as the latter refused to relinquish their economic and political sovereignty. John M'Cosh described some routes as 'tedious' but agreed that they were also proof of a significant volume of commercial transactions. 'There is

26 Kakati, *Assamese*, p. 380; Medhi, *Assamese Grammar*, p. 36-42.

27 Colebrooke, *Remarks*, p. 197.

28 Chevalier, *Adventures*.

29 Bhuyan, *Anglo-Assamese Relations*; Choudhury, 'British quest', p. 180.

30 Lindsay, *Anecdotes*, p. 174.

31 Wood, *Map*; Bose, 'Some Account'.

32 Letter from Secretary to the Chief Commissioner, Assam to Secretary to the Government of India, Foreign Department, no. 1194, 21 June 1886, Shillong quoted in Allen, *Assam District Gazetteer*, p. 82-83.

an open road from Upper Assam into Burma and thence into China by which a considerable trade in Chinese and Burmese manufactures is carried on.'[33] His descriptions pointed towards the possibilities of reframing the India-China trade and the opportunities that these expansive mountainous spaces could offer to the EIC.

By the 1850s, a wealth of detailed reports was published that included routes, human activities, and the large volume of commerce.[34] These geographical descriptions embodied new materiality, possibilities of private commerce, romanticized ideas about the landscape, and increasing Anglo-China political rivalries. The EIC officials understood that continuous flows of commerce, human mobility, and political patronage from the rulers had kept these routes dynamic. The accounts were fortified with tools of commerce and maps with accurate details of trading joints, markets, or distances etc.[35] These routes were at the root of a dynamic environment for increased flows of commodities, colonial military, subjects, traders, animal wealth, and capital.[36]

The EIC tried to break into highly controlled communication routes but were initially unsuccessful. Early commercial transactions were often marred by trade disputes with the residents of the eastern Himalayas. In 1826, David Scott, EIC's top official in Assam, spent a sum of 900 rupees in sending out a consignment of goods.[37] The consignment was 'plundered on the way'.[38] But the junior EIC officials did not lose hope.[39] Unsure of sovereign control of the highlands, they pushed for establishing markets in the foothills. The 'hill tribes are dependent on Assam in great measure for the necessities of life for which and the few articles of luxury they indulge in they carry on a useful barter'.[40] These officials were swayed by the idea of the prospects of bringing together the eastern Himalayas under one single political and economic unit. Of the list of possibly exportable goods, an official recommended red bread-cloth, white cotton cloth, muslin, coloured handkerchiefs, velvets, and castor oil, while the list of imported goods

33 M'Cosh, *Topography*, p. 11-12.

34 Hay, *Map*; Cooper, *New Routes*; Jenkins, 'Note', p. 251-254; Peal, 'Note', p. 69-82.

35 Williamson, 'Lohit-Brahmaputra', 364.

36 Cederlöf, *Founding an Empire*; Iqbal, 'Space', pp. 69-84.

37 Secret and Political Consultations, Bengal, 1827, 7 September, No. 17-19, quoted in Barpujari, 'Attempted Traffic', pp. 326-331.

38 Ibid.

39 Bose, *British Policy*, 75.

40 'Letter from F. Jenkins, agent to the Governor General to Government of India, no. 12, 15 February 1844', quoted in Dutta, *Hand Book*, p. 27.

could comprise ivory, gold, amber, copper, *daws*, and spears.[41] The Calcutta Chamber of Commerce, the influential British traders' lobby, proposed to market Assam tea in Tibet. They hoped that these tracts would not come under the political control of the Tibetans and Chinese. Francis Jenkins, an EIC official, hoped for an uninterrupted trade route to Tibet and China through these complex geographies.[42] But this was easier said than done. Meanwhile, as these routes acquired new political and economic meanings, there was much more excitement in the area.

Plants, commodities and commerce

As the EIC merchants began their search for mountainous routes, the eastern Himalayas brought new commercial prospects with its forest products. The English officials noticed how the mountain residents were also involved in commercial activities. 'Mishmis are constantly on the move in their trading expeditions'.[43] The EIC's desire to merchandise a rich biodiverse landscape bore fruit. Of many products that offered commercial possibilities, the Mishmi Teeta (*Coptis Teeta*), a widely used medicinal herb, was already part of the pre-company-era merchandise.[44] It was widely used in Indian pharmacopeia and Chinese medical practices.[45] China's Szechwan exported Mishmi Teeta as a popular traditional drug as *Huang lien* or *shui lien*.[46] Dalton, a British anthropologist in India, praised the herb for bringing wealth to the mountain dwellers. 'The chief sources of wealth to the Mishimis are ... the valuable medicinal plant 'Coptis teeta' or Mishmi teta'.[47]

In the 1820s, Nathaniel Wallich, the Dutch surgeon, botanist, and superintendent of the EIC's Botanical Garden in Calcutta, noted that its 'root constitutes a drug known among the Mishmees and Lamas in the mountainous regions bordering upon Upper Assam' and that it was widely familiar to the residents of both the eastern Himalaya and China. 'Amongst

41 'Letter from Neufville to David Scott, 27 January 1830, Political consultations, Bengal, Nos. 29-30, 14 May, 183', quoted in Barpujari, 'Attempted Traffic', p. 327.

42 Ibid. 27-28.

43 Dalton, *Descriptive Ethnology*, 16.

44 Watt, *Dictionary*. I acknowledge Santosh Hasnu for drawing attention to this plant during the Spring School, 'Modern Empires, Flows, Environments and Livelihoods' held in March 2017 at IIT Guwahati.

45 Long and Huang, 'Coptis teeta-based', pp. 343-349; Long and Huang, 'Traditional cultivation', pp. 79-86.

46 Henry, 'Chinese Names', p. 245.

47 Dalton, *Descriptive Ethnology*, p. 16.

these three nations it is in great estimation'.[48] Wallich agreed that it was 'in universal use as a powerful tonic and stomachic'. Assam's commissioner Francis Jenkins described how 'quantities are sent down to Assam in neat little baskets with open meshes made of narrow slips of rattan or some such material and measuring three to four inches in length by two and a half in breadth and one and a half in width'. Jenkins noted how 'each basket contains about an ounce of small pieces of the root from one to three inches long; they are nearly cylindric uneven scabrous more or less curved of a greyish brown colour varying in thickness from the size of a crow quill to double that diameter.' The dried root of this plant arrived in Bengal in 'small rattan baskets each containing from one to two ounces'. By 1850s, the Mishmi Teeta was an integral part of the China-Eastern-Himalayas-Assam, and European commercial flows. Meanwhile, this had become part of *Pharmacopoeia of India*.[49] The Mishmi Teeta, imported as a medicinal drug into Bengal, was also on display at the Royal Botanic Gardens, Kew, as of the 1880s.[50] In the 1910s, an *ad valorem* tax was imposed on this product.

The search for other forest products continued. Further south, the upland investors tried to expand agrarian activities into the hills and faced strong resistance from the hills' residents.[51] Vowing not to make the same mistake in the eastern Himalayas, the English, and also now joined by the Indian merchants, looked to aggressively merchandise the eastern Himalayas. The best example of this was many 'sorts of gums' which were already part of the commercial flows into Bengal. As early as in 1810, the EIC officials reported on the presence of the India rubber, *Ficus elastica*, across the eastern Himalayan area and northern Burma. Accounts of the botanical aspects of the plant were produced in the 1830s.[52] Though this gum had competitors, William Griffith, an EIC botanist, deemed it superior: 'If strength, elasticity, clearness and freedom from viscidity as well as from foreign matter be [a] test of excellence then this product may be considered superior to any other hitherto manufactured'. For the next half century *Ficus elastica*, in high demand in Britain, thrived commercially. Ranking high on the list of merchandise imported from the mountains, it changed hands across diverse economies. The India rubber, or *Bar Attah* in the vernacular world, requiring many layers of physical work, was skilfully

48 Wallich, 'Notes', p. 347; Wallich, 'Medico-Botanical Notices', p. 196.

49 Waring, *Pharmacopoeia*, p. 4-5.

50 Kew Royal Botanical Garden, *Official Guide*, p. 7; Assam, *Report on the Trade Between Assam and the adjoining Foreign Tribes and Countries, 1924-25*, p. 2.

51 Ludden, 'Cowry Country'.

52 Griffith, 'Report', p. 137.

collected from the eastern Himalayas and Burma, before being carried to the foothills.[53] 'Much of the rubber does not come from within' the British 'revenue limits; much comes from the tracts inhabited by tribes in a state of semi-subjection and more still from the country which is not under subjection at all'.[54]

The first shipment containing this gum reached London in 1847. This initial venture met with a temporary glitch. Initially, this trade remained in the hands of the English traders.[55] Two English trading firms procured trading rights in the mountains.[56] Soon, Indian traders – mostly the Marwari traders from India's desert state of Rajasthan who followed in the footsteps of EIC traders and invested in the lowlands of Assam – expanded their operations into this trade. One of them was Haribilash Agarwalla, who recounted that in the 1860s his father, who had arrived in Assam only decades earlier, became part of this trade. Haribilash's father negotiated with residents of the mountains to supply India rubber.[57] In the 1870s, Haribilash, whose mercantile activities included trade in agricultural and forest products, entered into an agreement with a chief in the mountain for annual supplies of India rubber. Over the next couple of decades, Haribilash earned a handsome profit from this trade, which allowed him to undertake other proto-capitalist ventures.

With increased speculation in this trade, business rivalries intensified, resulting in increasing government regulation.[58] 'India rubber trade is a subject of great complication all over the frontier'.[59] This regulatory increase allowed the colonial government to move into the mountains, which made mountain residents unhappy. 'The Singphos and Khamptis are a little sore about the prohibition to cut rubber when and where they like', wrote a colonial official in 1873. Nevertheless, throughout the 1870s, trade in India rubber was the highest source of revenue for Assam's trade with the eastern Himalayas.[60] Britain received as much as 355,616 kg of India rubber in 1876.[61]

53 Wardle, *Empire*, p. 143.

54 Bengal, *Report, 1871-72*, p. 13.

55 Goswami, *Aspects*, p. 123-124.

56 Hunter, *Statistical Account*, p. 141.

57 Agarwalla, *Haribilash Agarwalla*, p. 24.

58 Bengal, *Report, 1871-72*, p. 310; Assam, *Annual Report on the Administration of Land Revenue in Assam, 1874-75*, p. 45; Kar, 'Historia Elastica', p. 131-150.

59 Bengal, *Report, 1871-72*, p. 13.

60 Assam, *Report on the Administration of the Province of Assam, 1878-79*, p. 125.

61 Baynes, *Encyclopedia Britannica*, p. 839.

Tighter regulation did not stop trading firms like that of Haribilash from remaining in this commercial venture. When a stock of this product was confiscated, Haribilash bribed a colonial official to release his impounded stockpile of rubber.[62] A criminal case was lodged against his firm, leading to acrimonious legal proceedings. The firm's stake in this trade was much higher than such official regulation. Haribilash hired a Calcutta-based English lawyer to save him from the criminal proceedings against him.[63] Speculative trade in India rubber further intensified trade rivalries amongst the shareholders of the Indian Marwari trading firms, between the Marwari and the British private traders, and among the hill residents. 'The Singphos and Khamptis ... are jealous of the money made by its collection by the Mishimis'.[64] The colonial restrictions did not stop more firms from entering into this trade either. By the turn of the century, these traders were found 'in the midst of the jungle, on the paths leading to the mountains from which the wild tribes come; and it is exclusively with him that these visitors do business'.[65] Their profits were also high. 'Kyahs make very large proportionate profits in their dealings with hillmen, making, for instance, cent. [sic. one hundred] per cent on rubber, or even more'.[66] The Marwari traders' success came through the establishment of social relations with the mountain residents and by purchasing mountain products from the hill collectors at a much cheaper rate.[67] Haribilash consolidated his commercial ventures all along the foothills of the eastern Himalayas into the 1890s. The commercial flows did not remain confined to forest products: artisanal production from specialized localities was part of the trans-frontier flows. Brass utensils produced by artisans on Assam's floodplains regularly reached foothill markets for further northward mobility.[68]

These re-energised commercial flows integrated residents of the mountains, the Indian Marwari traders, and the British firms – many having their offices in Calcutta – into one common economic unit. These commercial exchanges were dependent on both appropriation and defiance of state-sponsored connections. By the 1890s, the global expansion of the British Empire, aided by a set of complex techno-bureaucratic mechanisms, was

62 Ibid., p. 35.

63 Ibid., p. 37.

64 Bezbaruah, 'Rubber Dispute', p. 157.

65 Assam, *Physical and Political Geography of the Province of Assam*, p. 51.

66 Assam, *Report on the Trade Between Assam and the adjoining Foreign Countries for the year 1881-82*, p. 6.

67 Hunter, *Statistical Account*, p. 415.

68 Ibid., p. 141.

firmly entrenched in the foothills of the eastern Himalayas, but its command in the mountains remained insecure. To govern the seemingly uncontrollable commercial flows occurring further upland, the colonial government had taken other steps which ensured that both its command in the mountains as well as the flows of commerce progressed cautiously.

Redesigning flows: Fairs in the foothills

We have seen that by the 1850s, the English and Indian merchants were well entrenched in the networks of commercial flows between lowlands and mountains. As modern capital knocked at the door of the foothills, and mountains' economic activities were still beyond its control, the EIC government tried to redesign these commercial links. As a part of this attempt, the well-established seasonal fairs were now made an integral element of the corporate governance. The fairs allowed for interaction of modern capital with a diverse set of economic exchanges under the supervision of the colonial officials.

In 1837, the Bengal government, which ruled Assam at that time, took steps to bring the diverse commercial transactions into a more organized event. An EIC-sponsored fair was organized at Sadiya, in the east of the eastern Himalayan foothills. By 1870s, these fairs appeared as state-sponsored events. The fairs were held across several weeks, normally spanning the months of February and March. Merchants from the highlands arrived in January, stayed in the facilities provided there until March, and returned thereafter.[69] The merchants – men, women, boys, and girls – came in large numbers, usually around 1,000-3,000 people. The list of merchandise carried by them, collected from faraway places and after crisscrossing varied environmental, political, and economic cultures, was only a moderate improvement on that of the previous centuries. The highlanders' merchandise included items procured from Tibetan and Chinese merchants in exchange for 'goods obtained' from these fairs.[70]

69 Assam, *Report on the Administration of the Province of Assam, 1876-7*, p. 6. 'Report on the Sudya Fair', Letter from Colonel Henry Hopkinson, Governor-General's Agent, North East Frontier, and Commissioner of Assam, to the Secretary to the Government of Bengal, Judicial (Political) Department, 6 March, 1873, in House of Commons Parliamentary Papers, Reports on Trade Routes and Fairs on the Northern Frontiers of India, (London, Her Majesty's Stationery Office, 1874), 52- 56.

70 Assam, *Report on the Administration of the Province of Assam, 1875-6*, p. 9; Ganguly, 'Modus Operandi', p. 16.

These commercial transactions were not insignificant in volume. The varied modes of transport that converged onto these sites of commercial transactions are proof of this. The lowlanders frequented these fairs in their bullock carts, widely used for transportation of peasants' produce on the Indian subcontinent, while the highlanders were dependent on mules and ponies for carrying their merchandise.[71]

Making these products available through a regulated market, and creating such annual markets, was not always solely the initiative of the British officials: the traditional chiefs of the mountains played an equal role. Not everyone could attend these fairs. The Udalguri fair, considered solely for the Tibetan Bhutias of Tawang, restricted 'inhabitants of Bhutan Proper'.[72] Not all fairs were economically or politically important. Some – like that of Udalguri or Sadiya – gained importance from commercial and political perspectives. The colonial officials regarded these two fairs as 'means of entering into communications with the tribes lying behind the Mishmies towards Thibet and China'.

The colonial officials and traders regularly attended the fairs. The government ensured police security as trade disputes ensued with violence. Activities included sports, horse races, and 'native games' by the merchants.[73] The fairs provided an opportunity for the colonial state to exchange gifts with the chiefs of the mountain dwellers. All this ensured a public showcasing of colonial superiority. Fairs allowed the colonial government to make further advances into the mountains.

Fairs also meant squeezing a longer spell of commercial exchanges into a short space of time. From the perspective of the government, this looked absolutely perfect in order to rein in these uncontrollable commercial exchanges, but such regulation was not always a success. In most instances, while the official fair took place for few days, the private commercial transactions continued beyond the limits imposed by the colonial government. Merchants continued to arrive even after notified dates. Non-conformity to the official dates was also due to other factors. Often a notification for a fair did not reach the mountain traders in time. These dates were also often simply not convenient for the merchant families.

These fairs, as part of the ritualisation of commercial transactions, largely took off as desired by the government. But things were regularly out of order. Fairs helped these pre-colonial commercial networks endowed with state

71 Ibid. p. 13.

72 Assam, *Report on the Administration of the Province of Assam, 1875-6*, p. 9.

73 Assam, *Report on the Administration of the Province of Assam, 1876-7*, p. 9.

patronages and 'maintaining some political communication'[74] with the residents of the uplands. This also helped to make products from diverse geographies become part of a larger commercial journey. Transactions that escaped the regulated networks of commerce were described as 'illegal trade' in colonial correspondence. In 1873, Henry Hopkinson, Assam's commissioner wrote to the Bengal government about this crisis. The government was 'suffering not inconsiderable loss and waste by the free trade in rubber … we are powerless to check'. Hopkinson mentioned how one of the 'last private steamers' that left Assam carrying merchandise from these foothills 'had as much as a thousand maunds of rubber on board'.[75]

As these fairs acquired new meanings after 1870s, a host of traders from different nationalities, including 'Bhutia' residents of Tawang, Chinese, European, Tibetan, and Indian Marwari traders, visited them. Haribilash frequented these fairs in search of new commercial opportunities.[76] The arrival of Marwari traders at these foothill fairs brought cash and new merchandise. These Indian traders provided credit to the highlanders for procuring the prized mountain produce at lower rates. Their tricks helped, and mountain products accelerated their way into the global market through Calcutta. Some fairs were solely devoted to the collection of India rubber. After decades of experience, these traders successfully moved deep into the Himalayan and Patkai ranges. By the early years of the next century, Chinese, Marwari, and Tibetan traders jostled for profit in the lucrative ruby trade in Burma.[77]

Controlling and sustaining flows

While fairs became symbolic of the British colonial authority in the eastern Himalayas, they also helped to inject Indian merchant capital deep into the mountains. Fairs helped to regulate the commercial flows across the lowlands and highlands. But the precolonial orders of economic and political

74 Hunter, *Statistical Account*, p. 385.

75 One Maund = 40 kilogrammes. Letter from Colonel Henry Hopkinson, Governor General's Agent North East Frontier, and Commissioner of Assam to the Secretary to the Government of Bengal, *Accounts and Papers*.

76 Agarwalla, *Atmajibani*, p. 34.

77 Letter from Major W. C. S. Clarke, Deputy Commissioner of Zillah Luckimpur, to the Personal Assistant to the Commissioner of Assam, Camp Sudya, 21 February 1873, *Report on the Sudya Fair*; in Great Britain, House of Commons, *Accounts and Papers of the House of Commons*, Vol. 49, 5 March to 7 August 1874; van Spengen, *Tibetan Border Worlds*, p. 184.

exchanges between the highlanders and lowland populations were not erased. Early company correspondence condemned *posa* as 'blackmail', though it was allowed to continue.[78] In the 1880s, Mackenzie rephrased this description as he thought it was being rather 'inaccurately' called blackmail.[79] The colonial state's attempt to integrate these areas into the imperial economy also continued. The EIC began to construct roads across the volatile floodplain landscapes in 1836.[80] Pressure to build transport infrastructure connecting highlands and lowlands remained a major preoccupation for the government. Better infrastructure was the key to the success of the new era of commercial flows. To make fairs successful, the 'first necessity is a safe road to the plain', wrote a colonial official.[81] By the turn of the century, the foothills were securely integrated with major transport arteries, which withstood torrential rain and floods and allowed nonstop flows of commerce. These infrastructures intersected old routes to Bengal and reinforced the old river transport systems. Between the 1910s and the 1930s, more infrastructure projects were undertaken in the eastern Himalayas.[82]

In the late 1860s, the colonial officials realized the need for strict regulation of these flows. The tipping point comprised the acrimonious trade disputes amongst the India rubber merchants.[83] Meanwhile, the British tea plantations rapidly moved into the foothills or the mountains.[84] The arrival of modern capital further restricted mountain dwellers' mobility across these commercial spaces, and they fiercely resisted this. Colonial officials began to piece together thoughts about regulating both these commercial flows and safeguarding the tea plantations. The officials had no intention of stopping the profitable commercial transactions nor of relinquishing their control over these economic activities. This resulted in the promulgation of the Bengal Eastern Frontier Regulation of 1873. The colonial government had to give previous sanction to its subjects to proceed beyond the foothills. The lowland populations were restricted from moving into the mountains and the descending flow of commerce and humans would be regulated through 'blockades'. All these measures would safeguard the interests of private capital in the lowlands. This bureaucratic intervention

78 'Notes by Scott on "Welsh's Report on Assam 1794"', in Mackenzie, *History*, p. 383.

79 Mackenzie, *History*, p. 21-22.

80 Barpujari, *Comprehensive History*, p. 292.

81 Assam, *Report on the Administration of the Province of Assam, 1876-7*, p. 9.

82 Hasnu, 'History', pp. 111-115.

83 Bengal, *Report, 1872-3*, p. 64.

84 Guha, 'Big Push', p. 199-221.

ensured the regulation of these extensive commercial flows across such diverse geography. This legal apparatus, seen by the colonial officials as a mechanism for 'peace and governance',[85] guaranteed that the larger social, cultural, and natural conditions in which these economic activities were conducted were not to be dislocated. It also restaged pre-colonial political control in the frontiers for controlling the flow of humans and commerce across lowlands and highlands.

The foothills soon witnessed the rise of a new form of governmentality.[86] Colonial correspondence from the Assam State Archives describes the intense mediations of the government in this process. But none of them stopped the merchants from defying these checks. They adopted a wide range of tricks to sustain the flows of commerce across the assorted landscapes. The Marwari traders, often bypassing the colonial government, directly entered into trade negotiations with the mountain dwellers to buy mountain products. Such trade agreements were regularly flouted, leading to disputes.[87] Understanding the inherent flaws, the government issued licences or 'passes' to the merchants to move into the areas of the eastern Himalayas merchants.[88] These passes acted as a double-edged economic instrument and soon metamorphosed into permanent shops, allowing the merchants to establish strong social and economic links with the mountain residents. Some commercial transactions, including the buying and selling of elephants, were regulated by the government.

Desire for modern transport

> Every limit appears as a barrier to overcome ... while capital must on one side strive to tear down every spatial barrier ... for its market.[89]

By the late nineteenth century, with the deep penetration of Indian and British merchant capital into the eastern Himalayas and the Patkai, the possibility of overcoming all obstacles to smoothen the commercial flows across these rugged landscapes began to attract broad attention. The possibilities of transforming those risky trading zones into the hub of modern transport,

85 Bengal, *Report, 1872-3*, p. 93.

86 Assam, *Report on the Administration of the Province of Assam for the years 1874-75 & 75-76*, pp. 3, 75-76.

87 Chakravorty, *British Relations*, p. 147.

88 Ibid., p. 64.

89 Marx, *Grundrisse*, p. 524.

either by taming the natural barriers, by modernizing, or by mechanizing the lethargic transport system, were regularly debated. Several English officials, including Arthur Cotton, the British engineer who built the massive irrigation system in India's southern provinces, proposed integrating these regions through comprehensive railway and riverine networks. Cotton proposed creating a waterway between the Brahmaputra and Irrawaddy as a conduit for flows of British and Chinese goods. A railway grid connecting these regions, imagined as a technological wonder of modern times, was also proposed. These proposals evoked passionate technical debates but did not garner enough favour from the colonial government and eventually had to be shelved. In the 1890s, the official dream of modernizing the India-Burma routes through the Patkai reached a new level of momentum. This sudden intensification of interests to re-energize these flows with much wider networks of a modern transport system – the Bay of Bengal on the one side and the Burma-China border in the other side of the rim – was due to several factors. One was the rise of vibrant border trading between China and Burma which had grown significantly after the British annexation of Upper Burma; an increase estimated at 30 per cent between 1896 and 1899.[90] The bordering areas of Assam, Tibet, and Burma acquired extraordinary commercial and industrial importance because of the gradual transformation of these areas into highly productive sites for production of tea, coal, petroleum, and timber.[91]

Several surveys designed to identify the potentiality of transport modernization were undertaken during this period. It was hoped that this modernization would consolidate cross-border trading across Southwest China and Bengal through Assam. Robert Atwell Way, a British engineer, was tasked with producing a report on the feasibility of connecting Assam and Burma through a railway system. Of the three possible routes that he proposed, the shortest and preferred one was the route that proposed to connect the north-eastern corner of Assam with Burma's Mogaung.[92] The government shelved the idea, seeing no significant commercial possibilities. Noel Williamson, writing in 1909 about the regions between Assam and Southeastern Tibet, lamented that the 'trade [here] is infinitesimal'.[93] 'The imports which pass up to Tibet from Assam through Miju traders amount to little, and of Tibetan exports there are none'. Williamson in fact asked whether 'these conditions' would have continued 'if an easy and fairly

90 Cotton, 'Communication', p. 255-259; Iqbal, 'Space', p. 74.

91 Saikia, *Forests.*

92 Way, *Assam-Burma*, p. 1-2.

93 Williamson, 'Lohit-Brahmaputra', p. 382.

expeditious route existed'. His mercantile instinct compelled him to note how 'thousands of maunds of wool are wasted annually simply because there is no market.' The only option was to have an improved line of communications along 'the natural outlet' and 'line of least resistance'. Williamson hoped that a 'good bridle path' from the 'limits of the British territory' to Sadiya which was close to a rail head, would radically transform the prospects of trade with Tibet. All these speculations about having a modern transport network resulted in nothing by the early twentieth century.

In 1905-6, E. C. Young, another British official, travelled on one of those routes from Yunnan to Assam. He came across 'a small trickle of trade' in 'Indian tea, blankets and matches'. Young was convinced of 'the existence or practicability of a direct trade route from China to India'.[94] 'The course of trade, like water, follows the line of least resistance', Young recounted. His difficulties were not those of routes but were only because of his lack of knowledge of the language. 'Many of our difficulties arose from ignorance of the language, as although I speak Chinese and Hindustani, neither of those languages were of use much beyond the frontiers of the two empires'. His route passed through a 'main trade route' which had been 'used by the Chinese traders as well as by the inhabitants of the country'. Here lies a great prospect of commerce which could be materialized by 'the construction of lines of communication, such as railways and roads'. 'The chief difficulties of this forest tract are the lack of roads and supplies, and heavy rainfall, whilst the lofty ranges of mountains strictly limit the number of possible routes'. It was the mismatch between the economic logic ['large stocks of provisions'] and geographical conditions which constituted a hindrance.

Despite Young's own anxiety concerning these geographical obstacles, commerce did take place. Both 'parties of natives are in the habit of passing through to Assam and back from time to time'. In this competing environment, all possibilities of a grand railway project stood no chance of economic survival. 'Isolated railways and railway systems in the East lose half their value and significance if they are not projected with the ultimate idea of creating and developing commerce over large and favourable areas.'[95]

By now things began to change fast in the eastern Himalayas as the Chinese control of Tibet, 'Sinification' as described by many, had picked up speed. This, along with the redefining of the borders of the nation-states across the eastern Himalayas, had additional consequences. Religious pilgrimage into India's lowlands, a vital element of these cross-border

94 Young, 'Journey', p. 176; D'Orléans, 'Journey', p. 566-582.

95 Holdich, 'North-Eastern Frontier', p. 387.

commercial flows, had slowly thinned out with the increasing presence of the Chinese state in Tibet. The colonial government made a frantic attempt to assert control over this fragile ecology and its indomitable habitants.[96] What followed was a series of raids by the colonial government in the eastern Himalayas. The government began to look anxiously at the incorporation of these discreet geographies into the political systems of two nation-states. Things subsequently began to unfold quickly, severely disrupting the commercial flows.

The symbolical retreat of the empire

The colonial government's eager endeavours to decisively absorb the commercial flows of the eastern Himalayas into the empire's orbit began to unravel in the 1940s. The most unsettling episode that unnerved these cross-border exchanges was due to World War Two. Within weeks of the Japanese occupation of Burma in 1942, the vulnerability of the British Empire's eastern frontier became clear. As chaos descended on Burma, the Indian and European population faced serious challenges from both the Japanese and the Burmese, and they streamed out of Burma. The early ones, mostly Europeans and a few rich and lucky Indians, managed to obtain berths on ships and seats on planes to travel to India. Unlucky ones in their several hundred thousands followed the routes in the Patkai, through which a network of modern transport infrastructure was the dream.

Among the evacuees was Purna Kanta Buragohain, an Assamese trader who had been in Burma for a decade when war broke out in Rangoon. Buragohain, who was at a China-Burma border outpost, returned to India. His diary gives a graphic account of his negotiations with local communities in Burma for a protected passage to Manipur.[97] Unlike the majority of evacuees, his journey was not through an unfamiliar landscape. His route, overlaid on the existing networks of flows, was dotted with sparse settlements whose inhabitants spoke languages that he shared, and with whose culinary practices he was familiar.

The retreat of the refugees from Burma was largely symbolic of the retreat of the British Empire, and the eastern Himalayas gained new momentum. Roads between Assam's foothills and China, despite the odds of nature, were built through investment of great sums of wealth and human labour.

96 Guyot-Réchard, *Shadow States*, p. 31-57; Hamilton, *In Abor Jungles*.
97 Buragohain, *Patkair Sipare No Bochar*, p. 234-254.

The enigmatic geographical barrier between Assam and China, which many wanted to dismantle in the previous decades, was now broken in a short time when military planes flew over the eastern Himalayas into China from the newly established aerodromes in the foothills of Assam. These aircrafts did not carry goods that once formed part of the flows, but instead soldiers, arms, and weapons, probably signifying the arrival of a new era for this region. As the war ended, the British Empire also retreated from India. The commercial transactions that had recently gained momentum came to an end. Activities started that were related to laying down national borders and strengthening national security strategies. These processes stopped the trans-frontier commercial flows explained in this chapter.

References

Agarwalla, Haribilash. *Haribilash Agarwalla Dangariyar Atmajibani.* Guwahati: Tarunkumar Agarwalla, 1967.

Ahmad, Nisar. 'Assam- Bengal Trade in the Medieval Period: A Numismatic Perspective'. *Journal of the Economic and Social History of the Orient* 33, no. 2 (1990): p. 176.

Allen, B. C. *Assam District Gazetteer*, vol. 3, Lakhimpur. Shillong: Government Press, 1906.

Amrith, Sunil. *Crossing the Bay of Bengal: The Furies of Nature and the Fortunes of Migrants.* Cambridge, MA: Harvard University Press, 2015.

Assam, Government of. *Annual Report on the Administration of Land Revenue in Assam, 1874-75.* Shillong: Assam Secretariat Press, 1875.

Assam, Government of. *Physical and Political Geography of the Province of Assam.* Shillong: Assam Secretariat Press, 1896.

Assam, Government of. *Report on the Administration of the Province of Assam for the years 1874-75 & 75-76.* Shillong: Assam Secretariat Press, 1877.

Assam, Government of. *Report on the Administration of the Province of Assam, 1876-77.* Shillong: Assam Secretariat Press, 1877.

Assam, Government of. *Report on the Administration of the Province of Assam, 1878-79.* Shillong: Assam Secretariat Press, 1880.

Assam, Government of. *Report on the Trade between Assam and the Adjoining Foreign Countries for the year 1881-82* [1882].

Assam, Government of. *Report on the Trade between Assam and the Adjoining Foreign Tribes and Countries, 1924-25* [1925].

Barpujari, Heramba Kanta. 'The Attempted Traffic with the Chinese through the North-East Frontier (1826-58)'. *Proceedings of the Indian History Congress* 24 (1961): pp. 326–331.

Barpujari, Heramba Kanta. *The Comprehensive History of Assam*, vol. IV. Guwahati: Publication Board, Assam, 1992.

Baynes, Thomas Spencer. *Encyclopaedia Britannica: A Dictionary of Arts, Sciences, and General Literature, 9th edition*, vol. 12. Edinburgh: Adam and Charles Black, 1881.

Bengal, Government of. *Report on the Administration of Bengal, 1871-72*. Calcutta: Bengal Secretariat Press, 1872.

Bengal, Government of. *Report on the Administration of Bengal 1872-73*. Calcutta: Bengal Secretariat Press, 1873.

Bezbaruah, Ranju. 'Rubber Dispute of the Miripathar Daflas; British Attitude'. *Proceedings of North East India History Association* 14 (1993): pp. 157–162.

Bhuyan, Suryya Kumar. *Anglo-Assamese Relations, 1771-1826: A History of the Relations of Assam with the East India Company from 1771 to 1826*. Gauhati: Department of Historical and Antiquarian Studies, 1949.

Blurton, T. Richard. *Krishna in the Garden of Assam*. London: British Museum, 2016.

Bose, Kishen Kant. 'Some Account of the Country of Bhutan', Translated by David Scott. *Asiatick Researches* 15 (1825): pp. 128–137.

Bose, Manilal L. *British Policy in the North-East Frontier Agency*. New Delhi: Concept, 1979.

Buragohain, Purna Kanta. *Patkair Sipare No Bochar* (*9 Years Beyond the Patkai*). Dhemaki: Purbanchal Tai Sahitya Sonha, 1993.

Cederlöf, Gunnel. *Founding an Empire on India's North Eastern Frontiers, 1790-1840: Climate, Commerce, Polity*. New Delhi: Oxford University Press, 2014.

Cederlöf, Gunnel. 'Corridors, Networks, and Pathways in India's Colonial North-east', *Fourth Jayashree Roy Memorial Lecture*, mimeo, North East India Studies Programme, Jawaharlal Nehru University, New Delhi, 2015.

Cederlöf, Gunnel. 'Seeking China's Back Door: On English Handkerchiefs and Global Local Markets in the Early Nineteenth Century'. In *Trans-Himalayan Borderlands: Livelihoods, Territorialities, Modernities*, edited by D. Smyer Yü and J. Michaud, pp. 125–144. Amsterdam: Amsterdam University Press, 2017.

Chakravorty, Birendra Chandra. *British Relations with the Hill Tribes of Assam since 1858*. Calcutta: K.L. Firma, 1964.

Chatterji, Suniti Kumar. *The Place of Assam in the History and Civilisation of India*. Guwahati: University of Gauhati, 1955.

Chevalier, Jean-Baptiste. *The Adventures of Jean-Baptiste Chevalier in Eastern India, 1752-1765: Historical Memoir and Journal of Travels in Assam, Bengal, and Tibet*, Translated by Caroline Dutta-Baruah and Jean Deroche. Guwahati: LBS Publications, 2008.

Choudhury, Deba Prosad. 'British Quest for Trade Routes from Assam to eastern Tibet: 1771–1914'. *Journal of the Royal Society for Asian Affairs* 8, no.3 (1977): pp. 180–184.

Christian, John L. 'Trans-Burma Trade Routes to China'. *Pacific Affairs* 13, no. 2 (June 1940): pp. 173–191.

Colebrooke, Henry Thomas. *Remarks on the Husbandry and Internal Commerce of Bengal*. Calcutta: n. p., 1804.

Cooper, Thomas T. *New Routes for Commerce: An Account of a Journey Made in an Attempt to Penetrate Thibet from Assam to Open New Routes for Commerce*. London: Henry S. King and Co., 1873.

Cotton, Arthur. 'On Communication between India and China by Line of the Burhampooter'. *Proceedings of the Royal Geographical Society of London* 11, no. 6 (1866): pp. 255–259.

d'Orléans, Henri. 'A Journey from Tonkin by Tali-Fu to Assam'. *The Geographical Journal* 8, no. 6 (1896): pp. 566–582.

Dalton, Edward Tuite. *Descriptive Ethnology of Bengal. By Edward Tuite Dalton. Illustrated by Lithograph Portraits Copied from Photographs. Printed for the Government of Bengal, under the Direction of the Council of the Asiatic Society of Bengal*. London: British Library, Historical Print Editions, 2011.

Devi, Lakhi. *Ahom-Tribal Relations: A Political Study*. Guwahati: LBS, 1992.

Dutta, Keshab Narayan. *Landmarks of the Freedom Struggle*. Guwahati: LBS, 1958.

Dutta, Keshab Narayan. *A Hand Book to the Old Records of the Assam Secretariat*. Shillong: Assam Government Press, 1959.

Gait, Edward. *A History of Assam*. Calcutta: Thacker and Sprink, 1906.

Ganguly, Jayanta Bhusan. 'The Modus Operandi of Trade between the Eastern Himalayan Subregion and Assam in the Pre-Colonial and Colonial Periods'. In *Border Trade: North East India and Neighbouring Countries*, edited by G. Das and R.K. Purakayastha. New Delhi: Akansha Publishing House, 2000.

Giersch, Charles Patterson. 'Cotton, Copper and Caravans: Trade and the Transformation of Southwest China'. In *Chinese Circulations: Capital, Commodities, and Networks in Southeast Asia*, edited by E. Tagliacozzo and W-C. Chang, pp. 37–61. Durham, NC and London: Duke University Press, 2011.

Goswami, Shrutidev. *Aspects of Revenue Administration in Assam 1826-1874*. New Delhi: Mittal, 1987.

Granoff, Phyllis. 'Illuminating the Formless, God, King, and Devotion in Assamese Illustrated Manuscript'. In *Bhakti and Power: Debating India's Religion of the Heart*, edited by C. L. Novetzke, S. Sharma, and J. Stratton Hawley, p. 131. Seattle, WA: University of Washington Press, 2019.

Gribble, R. H. *Out of the Burma Night: Being the Story of a Fantastic Journey through the Wilderness of the Hukawng Valley and the Forest Clad Mountains of the Naga Tribes People at the Time of the Japanese Invasion of Burma*. Calcutta: Thacker Spink & Co., 1944.

Grierson, George Abraham. *Linguistic Survey of India, Indo-Aryan Family. Eastern Group. Specimens of the Bengali and Assamese Language*, vol. 5. Calcutta: Government Printing, 1903.

Griffith, William. 'Report on the Caoutchouc Tree of Assam made at the Request of Captain Jenkins Agent to the Governor General'. *Journal of the Asiatic Society of Bengal* 7 (1838): pp. 132–142.

Guha, Amalendu. 'A Big Push Without a Take-off: A Case-Study of Assam, 1871-1901'. *The Indian Economic and Social History Review* 5, no. 3 (1968): pp. 199–221.

Guha, Amalendu. 'The Medieval Economy of Assam'. In *The Cambridge Economic History of India*, edited by D. Kumar and T. Raychaudhuri. Cambridge: CUP, 1982.

Guha, Sumit, and Kenneth Pomeranz. 'Life and Energy'. In *What China and India Once Were: The Pasts That May Shape the Global Future*, edited by B. Elman and S. Pollock. New Delhi: Penguin, 2018.

Guite, Jangkhomang. *Against State, against History: Freedom, Resistance, and Statelessness in Upland Northeast India*. Delhi: Oxford University Press, 2019.

Guyot-Réchard, Bérénice. *Shadow States: India, China and the Himalayas, 1910-1962*. Cambridge: CUP, 2017.

Hamilton, Angus. *In Abor Jungles: Being an Account of the Abor Expedition, the Mishmi Mission and the Miri Mission*. London: Eveleigh Nash, 1912.

Hamilton, Francis. *An Account of Assam [First Compiled in 1807-14] Edited by Surya Kumar Bhuyan*. Guwahati: Department of Historical and Antiquarian Studies, 1940.

Hasnu, Santosh. 'A History of Transport in Northeast India, 1824-1945', PhD diss., University of Delhi, unpublished.

Hay, John Ogilvy. *A Map Shewing the Various Routes Connecting China with India and Europe through Burmah and Developing the Trade of Eastern Bengal, Burmah & China*. London: Edward Stanford, 1875.

Henry, Augustine. 'Chinese Names of Plants'. *Journal of the China Branch of the Royal Asiatic Society* 22–23 (1887): p. 245.

Holdich, Thomas Hungerford. 'The North-Eastern Frontier of India'. *Journal of the Royal Society of Arts* 60, no. 3092 (1912): p. 387.

House of Commons Parliamentary Papers, Reports on Trade Routes and Fairs on the Northern Frontiers of India, (London, Her Majesty's Stationery Office, 1874).

Hunter, William Wilson. *A Statistical Account of Assam*, vol. 1. London: Trubner, 1879.

Iqbal, Iftekhar. 'The Space between Nation and Empire: The Making and Unmaking of Eastern Bengal and Assam Province 1905-1911'. *Journal of Asian Studies* 74, no. 1 (2015): p. 74.

Jenkins. Henry Lionel. 'Note on the Burmese Route from Assam to the Hookoong Valley'. In *Selection of Papers Regarding the Hill Tracts between Assam and Burmah and the Upper Brahmaputra*, pp. 251–254. Calcutta: Bengal Secretariat Press, 1869–70.

Kakati, Banikanta. *Assamese: Its Formation and Development: A Scientific Treatise on the History and Philology of the Assamese Language*. Guwahati: LBS, 1987 [1941].

Kar, Bodhisattva. 'Historia Elastica: A Note on the Rubber Hunt in the North-Eastern Frontier of British India'. *Indian Historical Review* 36, no. 1 (June 2009): pp. 131–150.

Kar, Bodhisattva. 'Nomadic Capital and Speculative Tribes: A Culture of Contracts in the Northeastern Frontier of British India'. *The Indian Economic & Social History Review* 53, no. 1 (2016): pp. 41–67.

Kew Royal Botanic Gardens. *Official Guide to the Museums of Economic Botany, issue 1, Dicotyledons and Gymnosperms.* London: Her Majesty's Stationery Office, 1883.

Lindsay, Robert. 'Anecdotes of an Indian Life'. In *Lives of the Lindsays or A Memoir of the Houses of Crawford and Balcarres*, edited by Lord Lindsay. London: John Murray, 1849.

Long, Chunlin, and Ji Huang. 'Traditional Cultivation of Coptis Teeta and its Values in Biodiversity Conservation'. *Biodiversity Science* 14, no. 1 (2006): pp. 79–86.

Long, Chunlin, and Ji Huang. 'Coptis Teeta-Based Agroforestry System and Its Conservation Potential: A Case Study from Northwest Yunnan'. *AMBIO: A Journal of the Human Environment* 36, no. 4 (2007): pp. 343–349.

Ludden, David. 'The Process of Empire: Frontiers and Borderlands'. In *Tributary Empires in Global History*, edited by P. Fibiger Bang and C. A. Bayly, pp. 132–150. London: Palgrave Macmillan, 2011.

Ludden, David. 'Cowry Country: Mobile Space and Imperial Territory'. In *Asia Inside Out. Volume III, Itinerant People*, edited by E. Tagliacozza, H. F. Siu, and P. Perdue, pp. 75–100. Cambridge, MA: Harvard University Press, 2019.

M'Cosh, John. *Topography of Assam.* Calcutta: Bengal Military Orphan Press, 1837.

Mackenzie, Alexander. *History of the Relations of the Government with the Hill Tribes of the North-east Frontier of India.* Calcutta: Home Department Press, 1884.

Malekandathil, Pius. 'Circulation of People and Patterns of Maritime Migrations in the Indian Ocean during the Pre-modern Period'. In *Migrations in Medieval and Early Colonial India*, edited by Vijaya Ramaswamy, pp. 259–286. London: Routledge, 2016.

Marx, Karl. *Grundrisse: Foundations of the Critique of Political Economy.* London: Penguin, 1973.

Medhi, Kaliram. *Assamese Grammar and Origin of the Assamese Language.* Guwahati: Publication Board, Assam, 1988.

Peal, S. 'Note on the Old Burma Route over Patkai'. *Journal of the Asiatic Society of Bengal* XLVIII, no. II. (1879).

Pemberton, Robert Boileau. *Report on the Eastern Frontier of British India.* Calcutta: Baptist Mission Press, 1835.

Pommaret, Françoise. 'Ancient Trade Partners: Bhutan, Cooch Bihar and Assam' (17th to 19th centuries)' (2000), accessed 2 December 2021, http://himalaya.socanth.cam.ac.uk/collections/journals/jbs/pdf/JBS_02_01_02.pdf pp. 1–25.

Ray, Haraprasad. *Trade and Trade Routes Between India and China, C. 140 B.C.–A.D. 1500*. New Delhi: Progressive, 2003.

Richards, John F. 'The Finances of the East India Company in India, c. 1766-1859'. Working Papers No. 153/11, London School of Economics, 2011.

Robinson, William. *A Descriptive Account of Assam: With a Sketch of the Local Geography and a Concise History of the Tea-plant of Assam*. London: Ostell and Lepag, 1841.

Saikia, Arupjyoti. *Forests and Ecological History of Assam, 1826–2000*. New Delhi: Oxford University Press, 2011.

Saikia, Arupjyoti. *The Unquiet River: A Biography of the Brahmaputra*. New Delhi: Oxford University Press, 2019.

Shneiderman, Sara. 'Are the Central Himalayas in *Zomia*? Some Scholarly and Political Considerations Across Time and Space'. *Journal of Global History* 5, no. 2 (2010): pp. 289–312.

Smyer Yü, Dan. 'Introduction: Trans-Himalayas as Multistate Margins'. In *Trans-Himalayan Boundaries: Livelihoods, Territorialities, Modernities*, edited by J. Michaud and D. Smyer Yü, pp. 11–42. Amsterdam, Amsterdam University Press, 2017.

Van Schendel, Willem. *The Bengal Borderland: Beyond State and Nation in South Asia*. London: Anthem Press, 2004.

Van Spengen, Wim. *Tibetan Border Worlds: A Geohistorical Analysis of Trade and Traders*, London, Routledge, 2000.

Waddell, Laurence Austine. *The Buddhism of Tibet or Lamaism*. London: W.H. Allen & Co., 1895.

Wallich, Nathaniel. 'Notes on the Drug called Mishmi Teeta and Pucha Pat'. *Madras Journal of Literature and Science* 5 (1837): p. 347.

Wallich, Nathaniel. 'Medico-Botanical Notices'. *American Journal of Pharmacy* 9 (1838): p. 196.

Wardle, Thomas. *Empire of India: Special Catalogue of Exhibits by the Government of India and Private Exhibitors, Colonial and Indian Exhibition, 1886*. London: William Clowes and Sons, 1886.

Waring, Edward John. *Pharmacopoeia of India*. London: W.H. Allen, 1868.

Waterfield, Henry. *Memorandum on the Census of British India 1871-72*. London: Eyre and Spottiswoode, 1875.

Watt, George. *A Dictionary of the Economic Products of India*, vol. 2. Calcutta: Superintendent of Government Printing, 1889.

Way, R. A. *Assam-Burma Connection Railway Surveys, Hukong Valley Route: Report and Approximate Estimate, 1896*. Calcutta: Superintendent of Government Printing, 1896.

Williamson, Noël. 'The Lohit-Brahmaputra between Assam and South-Eastern Tibet, November 1907 to January 1908'. *The Geographical Journal* 34, no. 4 (1909): p. 364.
Wood, Thomas. *Map Showing Route from Gwahatty to Kolltah Koosy on the Borders of Bootan*, no. 1, in Private Papers of Francis Buchanan-Hamilton, IOR Mss Eur D. 97, Asia and Africa Collection, British Library, 1793.
Yang, Bin. 'Horses, Silver, and Cowries: Yunnan in Global Perspective'. *Journal of World History* 15, no. 3 (September 2004): pp. 281–332.
Young, E. C. 'A Journey from Yün-Nan to Assam'. *The Geographical Journal* 30, no. 2 (1907): pp. 152–180.

About the author

ARUPJYOTI SAIKIA, Professor of History, Department of Humanities and Social Sciences, Indian Institute of Technology Guwahati, India. He studies early modern and modern environmental and economic history and Assam's political history. His publications include *Forests and Ecological History of Assam* (2011), *A Century of Protests* (2014) and *The Unquiet River* (2019).

Mobilities Today

7 How to Interpret a Lynching?

Immigrant Flows, Ethnic Anxiety, and Sovereignty in Nagaland, Northeast India

Jelle J. P. Wouters

Abstract

This chapter invites readers to Nagaland (Northeast India) to reflect on the impulse that compelled thousands of Nagas to participate in a horrific public lynching of a perceived illegal immigrant. The inflow of 'ethnic strangers' is seen as challenging the Nagas' ethno-territorial sovereignty and reinvigorates local obsessions with notions of autochthony – an emotive affirmation of Naga origins, roots, soil, genes, semen, and blood as the prime criteria of rights, entitlements, and belonging. This chapter presents immigration panic and ethnic violence as cultural and bodily friction blocking such flows. The implication? A body politic gripped by a volatile, potentially violent split between those considered autochthonous and those deemed outsiders.

Keywords: flow, friction, mob violence, immigration, ethnopolitics, Nagaland

5 March 2015, Dimapur, Nagaland, Northeast India

A fiery mob of thousands storm the Central Jail. They carry a blurry photograph, printed in local dailies the previous day, of a man accused of raping a Naga student (he did not) and of him being an illegal immigrant from Bangladesh (he was not). Police on duty are overpowered as the mob breaks open prison cells, one by one, allowing hardened Naga criminals to escape, until they find the man on their photograph.

The days leading up to this moment had witnessed massive outpourings of anger on the streets of Dimapur. Protestors carried banners and shouted slogans:

Cederlöf, Gunnel, and Willem van Schendel (eds), *Flows and Frictions in Trans-Himalayan Spaces: Histories of Networking and Border Crossing*. Taylor & Francis Group 2022
DOI: 10.5117/9789463724371_CH07

'IBIs [Illegal Bangladeshi Immigrants] get out of Nagaland!'
'Nagas are not safe in Nagaland.'
'This is our land, enough is enough.'
The commanding Naga Students Federation (NSF), meanwhile, issued the following directive: 'Unless we act tough on these people [Bangladeshi Immigrants], slowly but surely these people become masters of our Nagas in our own land.'
Now entering his cell, they beat Farid Khan, the accused, not ceasing their blows before he breathes blood. He is dragged outside. Then the real horror unfolds. He is stripped of his clothes, exposing to public view the weapon of his alleged crime: his phallus. With fists, sticks, and stones, the mob mutilates his body, turning his face pulpy.
A rope is then tied to his waist and strung to a motorcycle. In this way, he is paraded – naked and helpless – in the direction of the Clock Tower, a Dimapur landmark, where they intend to hang him. Along the way, a still swelling crowd – reports estimate between 7,000 and 10,000 Nagas – continue the torture, jeers, and cheers, and take photos and record videos, which are later shared on social media. By the time the procession reaches the Clock Tower, 7.5 kilometres of bodily brutality later, life has already left Farid Khan.
Still they decide to hang him.

The horrific events of 5 March 2015 are now some time behind us. Perhaps sufficiently so for us to begin to filter through the deeper politics, persuasions, and sentiments that generated the public lynching of Farid Khan on a Dimapur afternoon, in the clear light of day. The most straightforward analysis is to frame the lynching as vigilantism and popular justice, and then to subscribe to the long unoriginal conclusion that in India, as in many postcolonial settings, the state does not hold a monopoly on violence in the Weberian mould.[1] Relatedly, it can then be asserted that Nagas taking the law into their own hands is both evidence of state failure and a critique of that state, an argument that runs through recent lynching theory in Latin America.[2]

There are convincing grounds to pursue such an argument. Ancestral Naga-lands (of which the Indian state of Nagaland is a part[3]) are home

1 Brass, *Theft*; Comaroff and Comaroff, *Law and Disorder*.

2 See Goldstein, 'In our own', pp. 22-43; Uribe, 'Dismembering', pp. 79-96; Risor, 'Twenty Hanging Dolls', pp. 465-485.

3 The Naga are a transborder community whose contiguous ancestral territory is today divided between India and Burma (Myanmar) and within India between Nagaland, Manipur, Assam, and Arunachal Pradesh.

to protracted armed conflict with Naga insurgent groups seeking the rolling back of Indian and Burma (Myanmar) state structures. Moreover, Nagaland State is described as an epitome of misrule, and its politicians and bureaucrats are widely questioned for their ineffectiveness, corruption, miscarriages of justice, and inability to safeguard and promote the 'Naga good.' Disenchantment with the state and a fraught state-society political sociality at local level were undoubtedly part of a general atmosphere that enabled the lynching, and I will return to this point in some detail further below. Yet, interpreting the public lynching as either confirming the hypothesis that the Weberian ideal of sovereignty vested in the state's monopoly on violence is just that: an ideal which typically falls short in actual practice, or presenting it as an antidote to formal law and therefore as state failure are ultimately reductionist.

It is analytically more useful to interrogate the impulse – an amplified rage, a beastly, grisly urge – that compelled thousands of ethnic tribal Nagas to participate in this spectacle of bodily violence, either as enactors of egregious violence or as spectators cheering them on. The alleged rape of a Naga woman triggered the violence, but what was the deeper, 'the real', reason behind the fury of the Naga mob? Why, for one, was it a rape accusation against an 'ethnic other', and not any other crime (of which there are countless reported in Dimapur) or a rape allegation against a fellow Naga (instances of which are also reported) that galvanized the Naga body politic? What, furthermore, was the public transcript its perpetuators sought to convey through the callous cruelty and calculated exhibitionism of a hapless body? (Why else parade Farid Khan through the town and nail his body to a centrally located landmark, there for everyone to see?)

'Atrocity cannot be its own explanation, violence cannot be allowed to speak for itself, for violence is not its own meaning. To be made thinkable it needs to be historicized', writes Mahmood Mamdani in the context of the Rwandan genocide.[4] He continues: 'We may agree that genocidal violence cannot be understood as rational; yet, we need to understand it as thinkable. Rather than run away from it, we need to realize that it is the "popularity" of the genocide that is its uniquely troubling aspect' (ibid.: 8). While a public lynching, for all its horror, remains many miles removed from a genocide, I suggest that we need to embark on a similar project to understand the spectacle of bodily violence in Dimapur; to explore, that is, what made the public lynching of Farid Khan 'thinkable' to a large Naga mob. True, not all those parts of the mob actually committed violence. Yet, as Amy Wood

4 Mamdani, *Victims*, pp. 228-229.

noted about past lynchings of 'blacks' in America's southern states, the act of witnessing a lynching – and in our case also photographing and filming it and sharing it on social media – is 'to play a public role in lending value and authority to the spectacle.'[5]

To understand what made the Dimapur lynching 'thinkable' is not to see it as a spontaneous combustion of communal outrage, but to contextualize it, which requires it to be considered with the backdrop, I argue, of prevailing economic and political conditions, the logic and sentiments of the now faltering Naga ethno-national project, and – closely related to both – a proliferating panic about immigration and belonging produced by the increased mobility of surrounding populations. More to the point, this mobility pertains to transnational Bangladeshi immigrant labourers, whose rising, and formally illegal, presence in Nagaland (and in places across highland Northeast India) is challenging the nature of Nagas' everyday ethno-territorial sovereignty and now foments a reinvigorated local obsession with notions of autochthony; an emotive affirmation, that is, of Naga origins, roots, soil, genes, semen, and blood as the prime criteria of rights, entitlements, and belonging.[6]

Thus while, as the introduction and varied contributions to this volume convincingly show, the history of the India-China corridor was significantly a history of flows, the unfolding politics of the present appears progressively (yet far from exclusively) preoccupied with rootedness and exclusivity that cause ethnic tribal friction, at least so in Northeast India.[7] Among the Naga today, flows and frictions are mediated by ethnic tribal alliances and animosities. Independent India's resolve to uphold colonial cartography and institutions fragmented the Naga across international and federal states. These boundaries are locally deemed artificial. Layers of ethnic tribal control readily facilitate movements of Naga peoples, and their goods and ideas, across, even as nascent state identities also appropriate and divide Naga identities.[8] From the facilitators of cross-border flows, these same layers of control transform into blockages when the 'flowing subjects' are not fellow Nagas, but incoming 'ethnic others.' In Nagaland, and across highland

5 Wood, *Lynching*, p. 4.

6 This speaks to a globalist dialectic in which increased connectivity and proclaimed cosmopolitanism spur an apparently contradictory impulse towards an increased preoccupation with exclusive local belonging; a dynamic that Tania Murray Li (Li, 'Ethnic Cleansing') captures as a current, deep, and global conjuncture of belonging. On the perils of this renewed obsession with local belonging, see Geschiere, *Perils*.

7 Baruah, 'Politics', pp. 69-83.

8 Wouters, 'Sovereignty', pp. 97-116.

Northeast India more widely, 'immigration panics', of which this chapter presents an overture, and ethnic violence is on the rise and manifest itself as forms of cultural and bodily friction blocking flows. The causes of this friction are varied, as this chapter will show, but they relate contemporary ethnicity attachments and anxiety over the loss of ethnic sovereignty as central to the capturing of the flow-friction dialectic.

Immigrant flows and ethnic tribal friction

As I began thinking about what exactly was at stake in the Dimapur lynching, I soon came across two widely experienced local fears, namely of ethnic impurity and ethnic emasculation which congealed into a discernible sense of ethnic tribal anxiety at local level. Highly gendered, the first fear evokes the purity of Naga womanhood. It refers to the Baptist-Christianity-inspired values of pre-marital chastity and expectations – at once intensely ethno-nationalistic and deeply patriarchal – that Naga women produce 'pure' progeny. This trope of ethnic purity is felt to be threatened by intermarriages, followed by the creation of mixed offspring, between Naga women and immigrant labourers, instances of which are reported as being on the rise. This has already culminated in the birth of a new tribe colloquially called the 'Sumia' (a fusion of Sumi Naga and 'Mia', a local derogatory term for Bangladeshi immigrant), a residual category set apart from the normal scheme of Naga classifications into clans and tribes.

The other is the fear of Nagas' ethnic emasculation and the voiding of constitutionally enshrined ethno-territorial sovereignty (of whose political genealogy and praxis more below) over an ancestral homeland through the arrival of hordes of non-local, non-tribal immigrants. This 'invasion' of 'strangers', in terms of a sociological category that denotes those who come today and stay tomorrow,[9] is experienced as potentially causing the perceived pollution, even erasure, of Nagas' everyday practical sovereignty over their culture, territory, identity, and belonging. In the contemporary historical moment, as I shall show, the mobility and inflow of the 'immigrant other' into Naga ancestral land are experienced as contravening the rationale of the long-standing Naga ethno-national project. It is these fears that are now producing politics that are volatile and violent (or have the potential to be so) as they confront deeply affective assertions of ethnic awareness, sovereignty, and exclusive rights and belonging.

9 Simmel, 'The Stranger', p. 402.

What Simmel called generic 'strangers', Nagas specify as 'IBIs' (Illegal Bangladeshi Immigrants). Seen through a wider historical lens, Assam and (undivided) Bengal share a long history of mobility with peoples moving to and fro for purposes of trade and commerce, political refuge and intrigue, and thieving and tax evasion. The historicity of these movements is attested by both Ahom *Buranjis* (court chronicles) and colonial reports.[10] During the post-colony, both the nature and volume of in-migration began to alter. Partition followed by the emergence, in 1971, of Bangladesh turned most of this longstanding mobility into transnational flows formally subject to state surveillance and control. Unlike the mobile landless Bengali peasants who, sponsored by the Bangladesh state, were relocated in the Chittagong Hills in a purported state attempt towards colonizing and civilizing these hills,[11] the transnational mobility of Bangladeshi settlers in Northeast India is formally illegal, even though such stigma, and in a context of porous borders, does not significantly impede the continuing transborder labour movement of these mobile, landless peasants.

Furthermore, most new arrivals in Assam were now no longer wealthy Bengali/Bangladeshi traders, businessmen, or government servants, but originated from Bangladesh's most marginalized, most vulnerable, and poorest sections of the populace, who moved northwards in search of livelihoods. And whereas during the pre-colonial and colonial epochs, these migrants primarily settled on the Assam plains, amid Assamese political

10 Popular, and academic, imaginings of precolonial highland Northeast India as an insulated, unruly place mistakenly project the colonial effect of insulating and cordoning off the highlands – after attempts, in undisguisable mercantile spirit, to turn its inhabitants and lands into investment and other revenue-generating opportunities had failed miserably (Wouters, 'Keeping the Hill Tribes') – into a deeper history. This ignores, and blatantly so, that during the precolonial epoch these highland tracts were also the source and corridors for overland trade in mineral wealth (e.g., gold, copper, limestone, and jade) and commercial articles (e.g., elephant tusks, silk, opium, and lumber) between southern China, Burma, and Bengal. Consequently, the highlands and market towns in the adjoining foothills and plains were quite the 'contact zone' as they linked communities and trade caravans, dynasties and chiefs, foreign merchants and private investors, and early and competing mercantile corporations, of which the East India Company (EIC) was initially one amongst several (Cederlöf, *Founding an Empire*).

11 See Siraj and Bal, 'Hunger', pp. 396-412. It was in fact Partition and the creation of Bangladesh that spurred Bengali farmers into the hills. Whereas earlier they readily travelled across state borders in search of cultivatable lands, Siraj and Bal (ibid., p. 409) argue, they 'became immobile because of different regimes of border formation in this previously borderless region. Their desperate feelings of entrapment by the international state borders, along with the actual scarcity they faced, lured many of these marginal peasants to participate in the state scheme and to migrate into this area [the Chittagong Hills] and to contribute towards the decrease in the area's original inhabitants.'

renaissance and resistance including several bouts of communal violence in recent decades they moved ever further and into the highlands in pursuit of the most menial of labour, often under the most trying of conditions. What remains evident in all of this, is a gradual shift in mobility with Bangladeshi immigrants increasingly opting for the hills, where they find both labour opportunities and politically restive ethnic tribal communities. While this tension between immigrant inflow, ethnic purity, and ethno-national sovereignty reveals itself very forcibly in Nagaland, and so in relation to the protracted Naga Movement for self-determination, it is also discernible in ethnic tribal societies across the wider region.

In a seminal essay, Arjun Appadurai explains morbid ethnic violence as the expression of rising forms of social uncertainty over identity engendered by the forces of globalization, and in particular by the rapidly growing migration movements (both forced and voluntary) that increasingly complicate links between citizenship, territory, and ethnic identification and belonging. Amid this uncertainty, he argues, the body can become 'a theater for the engagement of uncertainty' through vivisectionist violence that transforms social uncertainty into 'dead certainty' by first mutilating and then eliminating the ethnic other from the body politic.[12] Among Nagas, I will apprehend this social uncertainty in terms of a collectively embodied 'ethnic anxiety.' Townsend Middleton theorized anxiety, drawing on psychoanalytical traditions and in the context of contested belonging in Darjeeling, as an 'affective state' that is deeply rooted in body and time, and so 'a dimension of and potential for markedly agitated forms of life and politics.'[13]

Whereas Appadurai relates ethnic violence to uncertainty over identification, and Middleton relates embodied anxiety to belonging, or rather the lack thereof, I concentrate on anxiety that emerges from sustained immigrant inflows and with that a perceived pollution and emasculation of ethnic sovereignty within a state-sanctioned territorial homeland. In her classic argument about *Purity and Danger*, Mary Douglas dealt with uncertainty, non-belonging, and anxiety in terms of 'pollution' and 'category confusion', which, if experienced, can spur a 'positive effort to organize the environment' by purifying it of that which is perceived as 'matter out of place', that is dirt, or something that 'offends the order.'[14] Douglas' argument can be helpfully adapted to apply to contemporary Naga ethnic anxiety. Here, 'matter out of place' locally manifests itself as the figure of the illegal Bangladeshi

12 Appadurai, 'Death Certainty', p. 226.

13 Middleton, 'Anxious Belongings', p. 608.

14 Douglas, *Purity*, p. 2.

immigrant, while the order that he (there are very few female Bangladeshi immigrants in Nagaland) 'offends' is the 'ethnic order'.

I approach the locally desired restoration of this ethnic order in terms of expressions of 'ethnic sovereignty', by which I mean a form of popular sovereignty asserted by ethnic subjects – in ways that are variously constitutional and unconstitutional, formal and informal, violent and non-violent, legal and lethal – and in the name of protecting and policing exclusive ethnic tribal and territorial control and rights. By framing it thus, I draw on scholarship that shifts the ontological grounding of sovereignty away from exclusive focus on the territorial nation-state and towards issues of internal and dispersed expressions and contestations of sovereignty within states.[15] Nagas articulate their ethnic sovereignty as a 'natural right' (besides it being constitutionally enshrined, of which more below) justified by their distinct 'non-Indian' history; a 'right' that now translates itself, within the wider contemporary Indian nation-state, in a regime of ethno-territorial exclusivity and exceptionality, one that is experienced as under threat by sustained immigrant inflows.

In our case, the assertion of Naga ethnic sovereignty manifested itself through the communal capacity to exert lethal violence on the body of a local 'stranger', and so in defiance of the rule of law and judiciary institutions. Seen thus, the public lynching of Farid Khan was a particularly dramatic expression of Naga ethnic sovereignty. If indeed so, the lynching was not a moment of medieval, pre-political darkness or an irrational lapse of an angry mob out of control, but an act of absolute political clarity that deflected or transcended, even if only momentarily, a pervasive existential anxiety of a Naga category under stress.

The causes, consequences, and corollaries of Naga ethnic anxiety and the expression of ethnic sovereignty through a public lynching are idiosyncratic in their particulars. Yet similar ethnic anxieties – manifest, lingering, or brewing – and assertions of popular sovereignty are now apparent across Northeast India. Two bodies of scholarship on highland Northeast India that have hitherto rarely been brought into the same conversation offer a context for this contemporary condition. The first is scholarship that focuses on state-led 'neoliberal developmentalism' and 'connectivity infrastructure'[16], the Look (now 'Act') East Policy, and new capitalist conditions and connectivities,[17]

15 Hansen and Stepputat, *Sovereign Bodies*; Das and Poole, *Anthropology*.

16 Bhattacharya, 'Neoliberal developmentalism', pp. 169-191.

17 Wouters, 'Neoliberal Capitalism'.

including the capitalist appropriation of nature and its resources in the region.[18] Jointly, these developments envisage highland Northeast India no longer as a periphery but as a thriving and integrated network of (trans)national capital, and a hub of trade, transportation, and communication; in short, a region of flows, connectivity, investment, and opportunity. As a corollary, the perception of the highlands as barren and unproductive has changed. They are now perceived as an area of opportunity and financial speculation, leading, for one thing, to the influx of 'fortune seekers' that range from boom and bust investors, corporations, and government departments on the prowl for mineral resources, to low-skilled migrant labourers desperate to make ends meet.

These ongoing economic transformations and the sustained inflow of immigrant labourers coincide with another strand of scholarship that reveals the progressive political preoccupation by ethnic tribal communities in the region with the pursuit, protection, and policing of state-sanctioned, distinctly non-liberal regimes of exclusive 'ethno-territoriality'.[19] Ethno-territoriality refers to Northeast India's unique 'ethnic homeland model' of politics and governmentality that essentializes the ties between ethnicity and territory.[20] It promotes 'variegated citizenship'[21] that locally separates and elevates those deemed autochthonous from non-local 'strangers' (not just illegal immigrants, but also referring to residents with the same nationality but who do not ethnically 'belong' locally). In Nagaland, as in other 'ethnic states' and 'territorial councils' in the region, autochthony, not Indian citizenship, serves as the basic criterion for rights, employment reservations, access to state resources, and belonging.[22] The securing of such an ethnic homeland has become the pinnacle of successful ethno-politics within the region, spurred by 'a perception that the STs [Scheduled Tribes] in the states with the most comprehensive protective discrimination regimes and rules of exclusion have done well economically and have been relatively successful in insulating themselves from being swamped by immigrants.'[23]

18 Karlsson, *Unruly hills.*

19 Baruah, 'Politics', pp. 69-83.

20 See van Schendel, 'Dangers', pp. 19-43; Baruah, 'Citizens', pp. 44-66.

21 Ong, *Flexible Citizenship.*

22 Besides Nagaland, Mizoram, Meghalaya, and Arunachal Pradesh function as de facto ethnic states and were all carved out of colonial Assam. With the exception of Nagaland, all Northeastern states are further subdivided into territorial councils. The means and ends of the governance of these territorial councils is the prerogative of ethnically delimited populations, governed by an ethnic tribal political class whose perpetuation is guaranteed through reserved electoral seats. See Baruah, 'Politics'; van Schendel, 'Dangers'.

23 Baruah, 'Politics', p. 5.

Put differently, what defines the economic and political condition in contemporary highland Northeast India is the emergence, and I argue convergence, of new neoliberal investments and connectivities and a non-liberal intensification of politics of absolute and exclusionary ethno-territorial belonging and rights. These two 'emergences' congeal in what I see as characteristic of the contemporary historical moment in highland Northeast India: a dialectical tension between new neoliberal connectivities and especially the heightened inflow of immigrant labourers and settlers, and increasingly volatile ethno-politics, deeply rooted in ethnic anxiety and steered towards protecting, patrolling, and legislating exclusive ethnic and territorial rights.[24] The implication? A body politic that is gripped, more than ever before, by a volatile, always potentially violent, split between those considered autochthonous or indigenous and those deemed non-local or outsiders.

The story behind the public lynching of Farid Khan offers a 'thick' and multifaceted narrative into this dialectical interplay of new neoliberal connectivities, immigrant labour flows, ethnic friction, and territory within the specific historical situation of Northeast India.

The Indo-Naga conflict and a town of violence

The Naga Movement for the right to self-determination is complex, compounded, convoluted, and much written about.[25] For present purposes, a brief summary will suffice. For well over six decades, the Naga Movement – in its now different forms and factions – has pursued the political refusal of the Indian (and Myanmar) state, and this through the barrel of the gun. Its ultimate objective is to realize a sovereign Naga homeland, even as in recent years the Naga rebel vanguard has begun to talk about 'shared sovereignty' and a 'special federal relationship' with India, provided that this recognizes and institutionalizes the 'uniqueness' of Naga history and culture. The Naga armed resistance provoked a massive escalation of violence from the Indian state, the dense militarization of the Naga highlands, the enactment by the Indian state of draconian extra-constitutional laws that drove a wedge between law and justice, and unspeakable tragedies.[26]

24 See Wouters, 'Neoliberal Capitalism'.

25 See Wouters, *In the Shadows*; Horam, *Naga Insurgency*; Iralu, *Naga Saga*.

26 Kikon, 'Predicament', pp. 271-282.

As an envisaged (but failed) political compromise, the Central Government enacted Nagaland State in 1963 and bequeathed it a special constitutional amendment, namely Article 371A that grants substantial political autonomy and exclusive ethnic and territorial rights and provisions to those classified as autochthonous to the state.[27] While the financing and functioning of Nagaland State are ultimately tied to Delhi and the Indian dispensation, its special constitutional status allows it to run as a *de facto* Naga ethnic homeland.[28] In Nagaland, only autochthones enjoy full rights, qualify for public employment, are eligible to stand for elections (out of sixty state assembly seats, 59 are reserved for Nagaland Scheduled Tribes), and can own land (with the crucial exception of Dimapur, where such restrictions on the non-Naga ownership of land are relaxed). Non-locals, moreover, are subject to an Inner-Line regime and need a permit for inward travel (again, with the exception of Dimapur for which, until very recently, no such permit was required).[29] The creation of Nagaland State as an ethnic homeland, however, did not conciliate the Naga National Council (NNC), which first spearheaded the Naga Movement until it was surpassed in power and influence by the rival National Socialist Council of Nagalim (NSCN) in the early 1980s. Insurgency continued, as did counterinsurgency.

In 1997, a ceasefire was agreed between the Indian state and the National Socialist Council of Nagalim – Isak/Muivah (NSCN-IM), the largest faction within a by then already splintered NSCN-fold. In following years, rivalling NSCN factions signed separate ceasefire agreements. At the time of the lynching in 2015, and the time of writing in 2020, these ceasefires had not resulted in a permanent political settlement. Nor had they resulted in the cessation of violence. The violence is now, however, not predominantly between Indian and Naga armed forces but between rivalling Naga factions. Somewhat paradoxically, factionalism escalated after the ceasefire,

27 Article 371A reads: Notwithstanding anything in this Constitution, (a) no Act of Parliament in respect of (i) religious or social practices of the Nagas, (ii) Naga customary law and procedure, (iii) administration of civil and criminal justice involving decisions according to Naga customary law, (iv) ownership and transfer of land and its resources, shall apply to the State of Nagaland unless the Legislative Assembly of Nagaland by a resolution so decides.

28 Wouters, 'Difficult Decolonization', pp. 1-28. Only, it must be qualified, for Naga tribes within the state of Nagaland. Article 371A does not extend to Nagas residing in neighbouring states. Naga communities residing in Myanmar have acquired levels of self-determination with their territory being declared, by the Myanmar Government, as a self-administrative zone.

29 In 2019, the Inner-Line gate was moved from the outskirts of Dimapur on the way to the state capital of Kohima, to the main entry gate of the city and putatively so in an attempt to police illegal immigrants within Dimapur more effectively. On the origins and evolution of the Inner Line, see Kar, 'When was the postcolonial', pp. 39-80.

spreading death and despair.[30] In this process, and because of the 'excessive' regimes of taxes and collections that each Naga faction imposed on ordinary Naga men and women, ostensibly to finance their struggle, a general air of disillusionment and unpleasantness began to emerge between Naga factions and the Naga populace they claimed to represent.[31]

Most the factional violence and the bulk of taxation occurs in and around Dimapur where Naga factions house most of their cadres in designated ceasefire camps. 'Walking in Dimapur is a visceral experience of militarization and violence', writes Dolly Kikon. She explains:

> Within India, Dimapur is the only city where two ceasefire camps of rival Naga insurgents are located adjacent to Indian security camps and headquarters. Factional shootings and killings in the heart of the city, unclaimed bodies, and the presence of Indian armed soldiers in public spaces, are common.[32]

While long beset by volatility and violence, the past decades nevertheless witnessed Dimapur grow into a sprawling and bustling commercial area. Additionally, significant state attention and funds were directed to its urban development and to connectivity infrastructure, and Dimapur became the largest, most populated, and best-connected settlement within Nagaland. The first Special Economic Zone in Northeast India was created on Dimapur's outskirts and invited (trans)national investment by promising conditions conducive to neoliberal profit-making. All this turned Dimapur into a place associated with investment, commerce, and economic growth. These associations emerged despite simultaneous conceptions of Dimapur as a debased world of danger, disorder, extortion, and factional violence.

First things first, however. Especially since the 1960s, villagers belonging to different Naga tribes left their home villages, often to escape relentless counter-insurgency operations and army brutality, and sought refuge in Dimapur. They were joined by poor hopefuls from the neighbouring and nearby states of Assam, West-Bengal, and Bihar who braved Dimapur's political volatility in pursuit of business, employment, and income. These non-local, non-tribal arrivals soon also included illegal immigrants from Bangladesh, where endemic poverty, devastating floods, and acute land shortages had pushed them across the Indian border. As stated by Amarjeet

30 Panwar, 'From Nationalism', pp. 233-258.

31 Wouters, *In the Shadows*, 81-123.

32 Kikon, 'From the Heart', p. 38.

Singh: 'Illegal immigration has been taking place in Nagaland, especially in areas bordering Assam, since the early 1970s or even earlier, but it has picked up since the 1980s.'[33] What in part spurred the initial arrival of Bangladeshi immigrants was an apparent 'dislike of the locals for manual labour' added to a general 'labour shortage.'[34] The illegal immigrants' willingness to perform any menial labour, the modest pay-checks they satisfied themselves with, and their experienced docility soon made them indispensable to Dimapur's growing economy. It was in the locally unappealing niches and crevices of Dimapur where they eked out a living.

The town began to expand further: new apartment clusters proliferated to house the growing Naga middleclass who demanded secure, pleasant, and comfortable lives, while shacks and shanty colonies populated by immigrants emerged on whatever land lay vacant (land often deemed 'uninhabitable' by Nagas themselves, such as fluctuating riverbanks and spaces beneath flyovers). Over time, however, Dimapur began to attract more immigrants than it could handle or desired, putting increasing strain on its infrastructure. Most new arrivals were now thought to be Bangladeshi nationals – although the 'suspects' themselves would insist that their places of origin were in southern Assam; in areas bordering the Indo-Bangladesh border but not from across it. Land, housing, water, electricity, and employment all became scarce, and life increasingly turned into a desperate scramble for jobs and livelihoods. As a result, a more restive town started to emerge; not suddenly or overnight, but in ways that were nevertheless noticeable.

As new urban challenges began to take their toll on the Naga citizenry, the presence of immigrants became more visible, increasingly seen as a source of the intensifying imbrications of congestion, competition, and crime into everyday life. Their presence also became perceived as contravening the rationale of the Naga Movement for a sovereign, exclusive Naga homeland. A pro-Naga discourse of ethnic and territorial control of Dimapur (and Nagaland) became louder and started to glare with apprehension at a local economy that now largely functioned on immigrant labour. In discussing this predicament, Nagas came to distinguish between 'legal intruders', or those who hailed from places within India, and 'illegal intruders', primarily Bangladeshi immigrants. While the presence of the former – Marwaris, Assamese, and Biharis – was backed up by some (albeit locally resented) legality, in relation to Bangladeshi immigrants it was their 'illegality' and

33 Singh, *Study*, p. 20.

34 Ibid., p. 29.

'out-of-placeness' that became emphasized. Back in 2002, a newspaper editorial warned thus:

> There is no denying the fact, that in any Muslim religious day [referring to Bangladeshi settlers who largely follow Islam], at least half of the shops in Kohima and some 75 per cent in Dimapur remain closed. The point is that this is a clear indication of how much the migrants have been able to make an impact on trading.[35]

A few years later, another editorial went further still: 'It may not be too long before somebody with the power of numbers (population) demands a Union Territorial status in Dimapur. Its market areas already look like a mini-Bangladesh, albeit more lucrative'.[36] In this atmosphere, Naga businessmen, contractors, and landowners who allowed these 'illegal intruders' a source of livelihood and residence, and therefore protection, also began to be viewed with misgivings and be accused of putting Naga exclusivity, privilege, and dominance at risk. Ardent criticism was also directed at the Naga government, which seemed unable, or unwilling, to check illegal immigration, and directed at local politicians who were suspected of enrolling illegal immigrants on electoral lists to be used as votebanks.

About a month prior to the lynching, the Naga Students' Federation (NSF) stepped up its earlier state-wide campaign to 'flush out IBIs'. Their crusade, its president explained, was 'not against any community or citizens from India', but a movement 'against illegal immigrants'. He explained further: 'immigration from Bangladesh has become a serious threat to Naga society.'[37] A few years earlier, in Mokokchung Town, the urbanized headquarters of the Ao Naga, a new organization emerged that called itself 'Survival Nagaland' and set itself the task of 'cleansing' the town of illegal Bangladeshis. 'Quit notices' were served and retribution promised to those who defied them, causing 'about 3,000 Muslim businessmen, traders, and labourers ... to flee Mokokchung.'[38] In other places, municipal and village councils embarked on 'enumeration drives' to identify the names and addresses of Bangladeshi immigrants residing within their territorial jurisdictions. If this seems relatively innocuous, the pages of history reveal how identification and documentation often precede bouts of horrific

35 Cited in Singh, *Study*, pp. 23-24.
36 Cited in Singh, *Study*, p. 25.
37 'Drive to Flush out migrants'.
38 Dutta, 'Dimapur Lynching'.

ethnic violence with organized mobs carrying lists of names of those to be 'removed' from the body politic. In short, a discourse and political atmosphere that targeted Bangladeshi immigrants had been building up over several years.

The rape accusation brought all of this to a head.

The lynching of Farid Khan

The story of the alleged rape is one of several twists and turns. For one thing, post-lynching it turned out that Farid Khan did not hail from Bangladesh but was an Indian citizen from neighbouring Assam. Moreover, medical examination of the proclaimed victim did not reveal clear-cut evidence of rape,[39] while CCTV footage showed her accompanying the accused seemingly willingly into a hotel. As a result, the spotlight turned on the proclaimed victim, who was now suspected of having falsely framed Farid Khan, of suffering from 'loose morals', or, even worse, of being a prostitute. The idea of a 'Naga prostitute', amid orthodox Baptist Christianity, caused entirely different moral alarm. These complications now punctured what had earlier appeared as a clear-cut rape case. But all of this emerged only after the lynching.

It took a few days before the story of the alleged rape reached the public, but when it did it landed on a soil already fertile with resentment against Bangladeshi immigrants. Multiple Naga civil and tribal organizations immediately released public statements that were duly published in Nagaland dailies. These statements raised a number of grievances. The NSF, for instance, stated:

> Time and again Naga civil societies have raised concerns about the danger of harbouring Illegal Bangladeshi Immigrants (IBIs) in our own home, giving them the shelter and security ... It is because of our obstinate attitude and relaxed nature, such heinous crime is being committed by the IBIs without any hesitation ...[40]

The Naga Council Dimapur (NCD) and the Naga Women Hoho Dimapur (NWHD) released a joint statement: 'This heinous crime has only exposed our Naga weaknesses and unless all Nagas take responsibility to tackle the

39 'Dimapur Lynching'.
40 'Naga Council Dimapur'.

menace of unabated IBI influx and their stay here in the state, crimes against our women and daughters by these people will only increase.[41]

Two things stand out in the above (and many other) press releases. Firstly, the alleged rape committed by Farid Khan was not seen as an individual act but indicative of the behaviour and desires of *all* Bangladeshi immigrants. Secondly, such acts, committed by outsiders on Naga soil, revealed Nagas' weakness and that their inaction would render them complicit in the emasculation of their very existence as an ethnic community.[42] The efficacy of these statements published by Naga apex bodies can hardly be overestimated. When powerful voices use ethnicity for collective identification and for articulating ethno-national interests, Maksic diagnoses in a different context that it is 'unique for their potential to reverberate across the social field', affecting how people interpret who they are, what their condition is, and whom they should trust and fear.[43]

Following these statements, a large protest was organized in Dimapur. In single file, the protestors marched to the offices of the District Commissioner where they demanded the 'culprit' to be either awarded the immediate death penalty or to be handed over them. Police were called in to defuse the crowd. The protestors then turned their anger on shops and properties that they thought were owned by Bangladeshi immigrants. A newspaper report concluded: 'the incident [of the alleged rape] has added fuel to the ongoing campaign against migrants in the state, particularly in Dimapur.'[44] The Nagaland Government responded by promulgating Section 144 of the Penal Code, which decrees the gatherings of people as unlawful. An exception, however, is made for a 'peaceful rally' that is organized the next day by the Naga Students Federation (NSF) and the Western Sumi Hoho, the apex body of the tribe that the proclaimed victim belonged to.

It is this 'peaceful rally' that morphs into a vengeful mob.

41 'NSF against IBIs'.

42 This second conviction takes on particular significance in view of the large number of Naga (and Northeastern more generally) females who for reasons of education and employment, or drawn by the opportunity that India's new modernity offers, migrate to cosmopolitan Indian cities, where they are frequently subject to sexual harassment and worse. Whereas in Nagaland these females can often rely on traditional support networks for protection, these networks are less powerful (though far from absent) away from their home state. Wouters and Subba, 'Indian Face', pp. 126-140.

43 Maksic, *Ethnic Mobilization*, p. 4.

44 'Students Protest Rape'.

Naga 'uniqueness' and the 'impure other'

Over the past years, a great deal has been said – in every conceivable voice – about the Naga Movement. Here, I focus on the foundational, affective, and politically expressive anchor of 'Naga uniqueness' that first actuated, and still galvanizes, the Naga ethno-national project. It is my contention here that this conviction of 'Naga uniqueness', of a distinct but benign Naga genetic endowment, natural essence, and genius, has for some years now begun to descend into a volatile preoccupation with the patrolling of Naga ethnic purity and completeness. This descent simultaneously connotes an emergence; an emergence of what Appadurai calls a 'threatened majority' that mobilizes itself as 'predatory' in reaction to an experienced '*anxiety of incompleteness* about their sovereignty.[45] Amid this anomic condition, the presence of 'the Other' becomes viewed as an obstacle to the naturalness, wholeness, and purity of the nation, likely resulting in demonization and violence directed against a minority population. Appadurai concentrates his analysis at the level of the nation-state, i.e., Hutus versus Tutsis in Rwanda and Hindus versus Muslims in India. I, however, argue that a 'threatened identity' can equally emerge in ethnic spaces located at the geographical, political, and symbolic margins of the state.

The founding Naga national principle of 'uniqueness' was first, and eloquently, articulated by A. Z. Phizo, the supreme ideologue and prophet of Naga nationalism, who served as President of the Naga National Council (NNC) from 1950 until his death in 'self-exile' in England in 1990. In the late 1940s and early 1950s, prior to the outbreak of large-scale violence, Phizo travelled across the Naga highlands, successfully whipped up Naga nationalist sentiment, instilled fear against the forces of what he presented as the invasive 'Hindu Government,[46] and demanded that villages, clans, and tribes unite and mobilize in pursuit of Naga independence. Phizo called his political ideology 'Naga-ism.' He explained himself thus:

> 'I am not a politician in the sense the word is understood in the West': 'I am neither a capitalist nor a socialist. I am a Naga from inside out. Firstly, because there is no alternative, secondly because I am a son of the soil, thirdly because there is no better social, political, and economic system than the way a Naga lives.[47]

45 Appadurai, *Fear*, p. 41 (emphasis in original).

46 Elwin, *Nagaland*.

47 Cited in Steyn, *Zapuphizo*, p. 133.

Phizo emphasized that the Naga way of life, based on a history, culture, and psyche characteristic only of itself, was threatened by the advance of the Indian state and its vast proletarian masses. Phizo most clearly articulated his conception of Naga uniqueness in his famous 1951 plebiscite speech, and I therefore quote him at some length.

> Being a nation, the Nagas have their own distinct ways of manners and living; and it is quite possible that we think differently in many respects ... The Indian mind greatly differs from the way we judge things ... As it is, there is nothing in common between the Nagas and the Indians. The difference is too varied, the feeling is too deep, and the attitude is too wide and too malignant for the two nations ever to think to live together in peace much less to become 'Indian citizens.' The only way to live in peace is to live apart ... Above everything else, we want to be free as a distinct nation: and we shall be free.[48]

In the remainder of his speech Phizo eulogizes Naga traditions, values, social equality, and moral economy and contrasts them with what he sees as the evils of Indian society: widespread landlessness, prostitution, caste, female infanticide, and poverty. In Papori Bora's apt analysis: 'Phizo articulated an absolute difference between the Nagas and Indians, by using the narrative constituted by Indology and Indian nationalism, but by reversing it. In the Naga nationalist narrative, Nagaland emerges as the better civilization than Indian civilization, unlike nationalist and Indological discourse, which presumes the pre-existence and superiority of the category India.'[49] In distinguishing Nagas from Indians, Phizo also invoked race:

> Racially, Nagas belong to the Mongolian family while the Indians belong to entirely a different race of their own ... As for the question of race, the less we talk the better. It is an undeniable fact that the Nagas are not Indian. We distinctly and unmistakably belong to the great Mongolian family. Strictly speaking, the world has come to know that the question of nationality is not a question of racial purity of a people. Also, the most important thing to consider is not merely one of politics but it is rather a problem of biology and psychology. To live together in peace different people must have the same attitude and the same feeling: there must

48 Phizo, *Plebiscite Speech*.
49 Bora, 'Politics', p. 249.

> be tolerance. Between the Indians and the Nagas, I am sad to say, these are lacking.[50]

Phizo's application of 'race' is somewhat puzzling: while he rejects racial purity as a principle of nation-building (and so was well ahead of his time) he nevertheless asserted that Naga 'biology and psychology' are irreconcilable with the contours and culture of Indian civilization.

Although Phizo's politics and leadership have come under renewed scrutiny in recent years, his framing of the Naga as distinct and unique remained foundational to Naga ethnic assertion and inspired an expansionist body of Naga nationalist literature. And though the later NSCN denounced Phizo's leadership on many counts, they similarly projected the Naga as 'a people with their own unique history, culture, and identity ... A unique socio-political community that has existed since time-immemorial',[51] one that exists at odds with the historical and cultural stretch of *Bharat*.

When the Indian state and the NSCN-IM began negotiating a political settlement, Naga uniqueness became a central theme. In 2002, the Indian Government, in an apparent political gesture, formally and publicly recognized 'Nagas' unique history and culture' (in precisely these words). The notion of 'Naga uniqueness' again became the spirit of the 2015 'framework agreement' that was signed (without its contents being divulged to the public) by the NSCN-IM and the Indian Government as a preamble to the final political settlement still under construction. On this occasion, Prime Minister Modi spoke thus: 'I have travelled to Nagaland on many occasions. I have been deeply impressed by the rich and diverse culture and the *unique* way of life of the Naga people.'[52]

But whereas Naga uniqueness was foundational to, and long irrigated, the Naga ethno-national project and now lies at the heart of the political negotiations – the theoretical frame, if you will, of a peace-deal in the making – this conviction is simultaneously experienced as at odds with a rising anxiety about the resilience of the Naga ethnic identity amidst fast-changing economic and political realities. These forms of anxiety are multiple. One such anxiety comprises the escalating intra-Naga differences and divisions that reveal themselves in near-routine factional violence and intertribal flares that now overlie and undercut projections of a singularly distinct Naga ethnic identity. Or in the words of a former Naga rebel leader:

50 Phizo, *Plebiscite Speech*.

51 'NSCN-IM on Jamir's Article'.

52 Kikon, 'What is Unique', pp. 10-11 (my emphasis).

'tribalism overrides sovereignty in Naga society.'[53] These divides, as noted, are also part of a rising sense of popular discontentment and disillusionment with the contemporary Naga Movement, whose 'national workers', as cadres of Naga factions are called locally, are seen to be progressively engaged in collecting taxes, intimidation, and internal bickering, rather than in selflessly advancing the Naga political cause.[54]

This anxiety over the unity of the Naga ethnic identity and the concerns over the contemporary condition-of-being of the Naga Movement are compounded further by the now longstanding adaptation of the Naga into two distinct nation-states, i.e., India and Burma/Myanmar, and within India into four states with dissimilar historical trajectories and contemporary experiences of government, development, and institutional politics. Naga nationalists use the term *Nagalim* (Naga-land) to protest this political fragmentation of a contiguously Naga-inhabited area. Colloquially, however, a distinction is today often maintained between 'Burmese Nagas' and 'Indian Nagas', as well as between 'Nagaland Nagas', 'Manipur Nagas', 'Assam Nagas', and 'Arunachal Pradesh Nagas', suggesting that this territorial division has now impinged on the socio-political imagination.[55]

Yet another source of uncertainty, and an excruciating one to many Nagas, is the slow pace with which political negotiations between NSCN-IM and the Indian state are unfolding, which have now entered their third decade. The contents of the negotiations remain shrouded in secrecy, yet it is simultaneously no secret that complete Naga independence has long since been ruled out. The prospect of seeing territorial and political integration of Naga territory within India, which is also demanded by the NSCN, is staunchly objected to by neighbouring state governments. This, too, now seems increasingly unlikely to succeed. What, then, after sixty-five years of bloodshed, suffering, and sacrifice is still left for the Naga? Will all these decades of conflict and misery end up in vain? These are questions that trouble and depress many Nagas.

A further uncertainty, and our main focus here, is the sustained inflow of immigrants on to Naga soil, and of whom most are in Dimapur. The historical experience of Tripura is often invoked as a comparison. In Tripura, the autochthonous tribal population became the victim of waves of immigrant arrivals from Bangladesh that saw them reduced to a marginalized minority in their own ancestral lands. This is a manifest fear locally and gives rise to

53 Swu, *Hails*, p. 37.

54 Wouters, *In the Shadows*.

55 Wouters, 'Sovereignty', pp. 97-116.

an elemental anxiety that progressively personifies itself in the figure of the illegal immigrant. In this way, the illegal immigrant, as a category, remains to most Nagas akin to what 'the neighbour' is to Slavoj Žižek. 'Beneath the neighbour', he writes, 'there always lurks the unfathomable abyss of radical Otherness, of a monstrous Thing that cannot be 'gentrified.'[56] While Bangladeshi settlers have, in Nagaland, long been a social familiarity, the relation between Naga hosts and these immigrants has never transcended the fear that their distinct physical features, culture, and religion may turn them into a rival and enemy in a different situation and time. Take the above forms of ethnic anxieties together, and we find that the Naga category is permeated with an existential tension that is increasingly experienced as intolerable.

It is in this context, and harking back to Appadurai,[57] that I argue a new sort of Naga nationalism is beginning to emerge on the ground. It is a nationalism that is increasingly fractured by internal factional and tribal divides, by political frustrations, and by elemental anxieties. It is a nationalism that now no longer primarily identifies itself through a benign regard for its own unity and uniqueness but through rising 'predatory' tendencies fuelled by a disregard of 'the Other.' With most Nagas now realizing that the chances of achieving independence (and territorial integration) are both complicated and slim, and in view of the shattering and experienced downfall of the Naga Movement, 'the Other' is no longer the idea of India at large (as it was when the Naga Movement first sprang up) or even Indian nationals, with whom most Nagas have now accepted a common existence as Indian citizens. Instead, the idea progressively confines itself to the figure of 'the Other' as people who have come to reside amongst them. It is now the 'internal Other', especially the figure of the illegal Bangladeshi immigrant, whose increasingly visible presence is perceived and highlighted as threatening the only form of sovereignty that Nagas can still anticipate to protect: ethnic, territorial, and cultural dominance on ancestral lands, as constitutionally safeguarded within Nagaland, through Article 371A.

In simpler terms, this new Naga nationalism connotes the downfall from 'Naga uniqueness' (the quality of being the only one of its kind), intrinsically defined, to a volatile and predatory preoccupation with patrolling and policing Naga wholeness and purity (cleanness, clearness, and the freedom from adulteration and contamination) that is defined in opposition to the 'ethnic other.' This reveals itself in, among other things, a series of

56 Žižek, *Did Somebody*, pp. 163-164.

57 Appadurai, 'Death Certainty'; Appadurai, *Fear.*

circulating and demonizing stereotypes that many Nagas currently hold and express about Bangladeshi immigrants on their soil, and against whom they define themselves in opposite terms. Such stereotypes include Bangladeshi immigrants' 'sexual virulence, natural (almost genetic) proneness to criminality, being uncivilized to the extent of being inhumanly dirty or unhygienic, breeding faster than dogs.'[58] It is a rising discourse, in short, that views Bangladeshi settlers as 'dirt', 'parasitic', and 'vermin' that ought to be contained and prevented from polluting the Naga body politic.

I postulate that, in broad strokes, it is this descent from Naga 'uniqueness' into concerns about Naga wholeness, purity, and everyday ethnic sovereignty that made the lynching of Farid Khan – who became the embodiment of this perceived threat against Naga purity and dominance – 'thinkable', and that now fuels a still dangerously simmering anxiety and fear directed against Bangladeshi immigrants as a category. This is not the full story, however. The ultimate threat to Naga ethnic identity and sovereignty, as locally perceived, is not just the presence of 'strangers'. Not even their rising dominance in the Dimapur and wider Nagaland economy. What particularly causes an intolerable anxiety is the injection of non-Naga blood and genes into the Naga body politic, thereby 'polluting' its presumed natural essence. It is to this fear, which is another consequence of immigration, that I presently turn.

Defiling Naga womanhood

What Michael Taussig contends about the 'defacement' of the human body (as well as of a nation's flag, money, statue, or any other national symbol), is that a 'strange surplus of negative energy is likely to be aroused from within the defaced thing itself',[59] which can equally, if not more dramatically, apply to the perceived *defilement* of a nation's women by 'outsiders', particularly in the context of experienced ethnic anxiety. This 'negative energy', once released, can result into, returning to Mary Douglas, a 'positive effort' to organize and purify the environment in order to restore order.[60] Can we not read into the public lynching of Farid Khan – an act at once real and symbolic and communicative – such a 'positive effort' towards purifying, repairing, and restoring the Naga body politic, and the wider desired ethnic order?

58 Hussain, 'Nation's Honour'.

59 Taussig, *Defacement*, p 1.

60 Douglas, *Purity*.

While rape is not unknown to Naga society, and is dealt with sternly under customary law, such crimes, when they occur, do not generally inflate into large-scale outpourings of public anger and lethal violence. 'The question that should baffle everyone', a commentator remarked post-lynching, 'is how the [Naga] society differentiates between rape of women by men of their own community and by men of another community.'[61] The collective violence over a rape allegedly perpetuated by an illegal immigrant on a Naga woman must therefore be analysed differently. That Farid Khan, as a non-Naga, 'dared' to 'force' himself on a Naga woman, was, I argue, interpreted as a sign of impending doom, of illegal immigrants becoming bolder, of an unfolding inversion of ethnic hierarchies, and of them not just challenging Nagas' authority in the spheres of economy and demography, but also defiling the very purity of the Naga nation by invading its sacred sites: its women. As such, it was taken as further evidence of Nagas' emasculation. What set the alleged rape in Dimapur apart was therefore not the act itself, but the non-Naga genes that were involved, and the image of these being forcefully injected into the Naga body politic. And this significantly changed the perception of what the rape was about.

The aggressive injection of outside genes into the Naga body politic has a particularly dark local history as revealed by devastating narratives of Indian soldiers raping Naga women in villages, at times making their fathers, husbands, and brothers watch at gunpoint.[62] 'Shame was brought upon us', a Chakhesang Naga elder once told me:

> Today some in our village are too dark-skinned to be *pure* Nagas. They are the children of rape. Indian blood is running through their veins. Whenever we see them, we remember those horrible days. We feel sad for them too. But what could we have done? We were helpless.

'Those horrible days' here refers to the time of the full-blown Indo-Naga war, not to the current ceasefire, to acts committed by gun-wielding soldiers, not by an unarmed and supposedly marginal immigrant, to a time when Naga villagers felt helpless against waves of violence that rolled over their villages, quite unlike the present scenario in which the Indian army has, by and large, returned to its barracks. It is my contention here that a deep connection existed between these past rapes, which brought 'shame and helplessness' to the community, and the particularly violent response that

61 Hussain, 'Nation's Honour'.

62 Iralu, *Naga Saga*, p. 185.

the Dimapur alleged rape elicited. However, this time the Naga did not feel 'helpless' and saw a real possibility to 'restore' Naga honour through the act of revenge.[63]

These past experiences and lingering sentiments further contributed to the inflation of the alleged rape from an individual criminal act into something much larger: into an attack on the Naga female body, on all Naga women, on the Naga nation writ large. Several commentators pointed out, aptly so, that the communal outrage and the lynching of Farid Khan was never foremost about the rape of a Naga woman per se. It was about how 'an outsider, more so a "lowly" Illegal Bangladeshi Immigrant (or IBI), violated the Naga nation's honour reposited in the bodies of its women by raping one of its "daughters."'[64] Complicating this narrative further, another commentator stated thus:

> The Dimapur lynching was never about justice. It was about patriarchal power ... Women are regarded as the property of a community and what outrages a community is not rape but the temerity of the "outsider" who lays claim to another's community's "property" ... The community stands exposed as not being able to have protected the honour of its women and children, or so it thinks.[65]

In this reading, prevailing patriarchy is added to the already multifaceted equation. To be sure, equality between men and women has never been central to a dream of a free and just Naga society, and whatever the perpetuators of the lynching thought of themselves, it was more than just a few commentators who read in the communal outrage the same old patriarchy and the same old male chauvinism that they insist have long constrained Naga women.[66]

At this stage of the argument, it must be emphasized that Farid Khan was a long-time resident of Dimapur and was married to a Sumi Naga woman with whom he had a daughter. This made him not just a 'rapist', and an adulterer, but also – in the eyes of many Nagas – a polluting source of ethnic impurity. It also identified him as a 'Sumia', a term introduced above. For many Nagas, the image of the Sumia condenses several nightmares, not the least that of 'category confusion.'[67] It blurs boundaries and creates uncertainty about

63 On the centrality of revenge, as a moral principle and political force, in the Naga context see Tinyi, 'Headhunting Culture'.

64 Hussain, 'Nation's Honour'.

65 Roy, 'Don't Touch'.

66 Kikon, 'City of Sorrow'.

67 Douglas, *Purity*.

who is, and who is not, a Naga by birth and loyalty. This sentiment is fanned by a longstanding fear of the ultimate genetic and cultural 'devouring' of Nagas by neighbouring populations that greatly outnumber them. It is fear that goes back at least as far as A. Z. Phizo, who spoke thus: 'We want our culture unhampered in the way we like, without having to worry for the possible mixture of alien blood.'[68] It is the emergence of the 'Sumia' community, and how this creates a local perception of Naga blood being diluted, that Phizo foresaw and feared. Equally concerned by this development, several influential Naga organizations have issued general cautions against intermarriage. The Dimapur Naga Women Hoho, for instance, instructed: 'Naga families would also do well to learn that marrying off their daughters to IBIs or adopting them does not beget anything good.'[69] Palpably evident, in this reasoning, is that the feared admixture of 'alien blood' centres on Naga women in a context in which labour immigrants are near exclusively male. Considering prevailing patrilineal descent, the possible liaisons and marriages between immigrant males and Naga females are perceived as particularly threatening, more so compared to Naga men marrying ethnic outsiders, instances of which, are in any case significantly fewer.

Among contemporary Nagas, Dolly Kikon writes, 'the emphasis of ethnic purity has become an important political and moral marker to claim territorial rights.' She introduces Lulu, a Naga woman who married a man from Assam and faced the music for doing so from her relatives, clan members, and villagers who collectively disowned her: 'By marrying a man from a different ethnic group she had stepped outside the ring of purity. She was not only an impure woman: even worse, she gave birth to children of mixed blood and thus became an outsider.' In the Assam-Nagaland foothills, whose territory is disputed between both states and where Lulu resides, she is perceived as a threat: [she is seen as] '"a traitor"; someone who is a member of the community but seeks to destabilize the politics of ethnic mobilization for exclusive cultural identities and homelands.'[70]

The notion of purity, here, is of a dual nature: it refers to purity of progeny, a Naga woman's socially ascribed duty of producing offspring that is

68 Phizo, *Plebiscite Speech*.

69 See 'Lawless Christian Intolerance'. To be sure, by configuring the Naga female body as a metaphoric referent for the purity of the nation, the Naga are hardly an exception; in nationalist projects around the world, and particularly in times of elemental anxiety, the female body, its sexuality, and reproductive organs, readily cease to be a private matter, but become inscribed with national importance as the purity of a nation's women, and the patriarchal control over adjudicating this purity, become associated with the quality and future of the nation itself.

70 Kikon, 'Difficult Loves', p. 231.

unmistakably Naga, but it also refers to Baptist Christianity and moral norms of pre-marital chastity. In a way, Naga society experienced a sexual revolution in reverse. In the precolonial and colonial past, less reticence was practised between the sexes, and pre-marital intercourse was socially accepted, even expected; 'virginity wins no halo in the Naga heaven, but is regarded rather as a sin', wrote Christoph von Fürer-Haimendorf.[71] This changed radically after widespread conversion to a particularly conservative and puritanical brand of American Baptist Christianity, which altered the way Nagas, especially females, became socialized and instructed about chastity and morality. This novel socialization relied on a 'morality of profit and loss' which teaches that once a man penetrates a woman but does not marry her, the girl's life is 'destroyed.'[72]

This moral transformation did not just change perceptions of sexual intercourse, but also of sexual violence and rape. In the past, Charles Chasie says, '[rape] reflected badly on the man that he has become so desperate as to commit such a crime. So, the shame of rape was more on the man than the woman.'[73] Per contra, there is now a 'popular belief that women got "it" (sexual violence) because they asked for it.'[74] In the aftermath of the lynching, when holes emerged in the hitherto official narrative of Farid Khan having 'destroyed' the life of a Naga female, the onus came to lie on the proclaimed victim, who now became variously shamed as a 'sex-worker' and 'a college girl looking for some fun', and the narrative changed to that of 'a consensual sexual rendezvous gone bad.'[75] The possibility that this connoted, of a Naga female voluntarily giving up her 'chastity' and 'purity' to an illegal Bangladeshi immigrant and so willingly defying Baptist teachings, produced an entirely different moral panic, but one that surfaced only in the aftermath of the lynching.

What emerges centrally in all of the above is a contemporary Naga nationalism that is also permeated by a powerful and patriarchal sense of 'uterine nationalism' that stems 'out of the recesses of the womb.'[76] Put differently, there is seemingly a societal conviction that Naga women should

71 Fürer-Haimendorf, *Naked Nagas*.

72 Kikon, *Life and Dignity*, pp. 26-27.

73 Cited in 'Naga Mob was Wrong'.

74 Kikon, *Life and Dignity*, p. 26. Kikon (ibid., p. 16) explains further that because of such conceptions, those Naga women who have experienced sexual violence tended to suffer this 'quietly', afraid of the 'social stigma attached to sexual violence and how the victims were usually held accountable for the actions of the perpetrators.'

75 Kikon, *Life and Dignity*, p. 16.

76 Heng and Devan, 'State Fatherhood', p. 349.

serve the Naga nation, not by taking up arms and disappearing into the jungle, but by protecting it from the 'pollution' of outside genes. It was the polluting of the 'Naga womb' and of the purity of its natural essence on two counts – through marriage and rape – that, in this reading, turned Farid Khan (and the category of illegal Bangladeshi immigrants he came to embody) into an all too perceptible threat against the purity of the Naga ethnic identity; a threat that needed to be removed.

Lynching as public sacrifice? The restoration of the ethnic order

Given the experienced ethnic anxiety, was the lynching of Farid Khan a rough and ready manifestation of a public sacrifice, in the way so influentially theorized by Rene Girard, in which a community forms and reasserts itself around an act of collective violence against a 'surrogate victim?' Girard elaborates that when internal rivalries, tensions, and violence threaten the survival of a community, this violence must be deflected outside so as not to destroy the community from within. A psychosocial mechanism arises – later dubbed the scapegoat mechanism – that leads to communal violence being projected upon a single individual who, by definition, is marginal or a stranger to the society and who, even if not in fact guilty of the community's problems and internal conundrums, is nevertheless treated as guilty by the crowd, which vents long-bottled-up fury and frustration. People who were formerly divided and struggling now unite against the selected scapegoat and rediscover their unity and sense of community by collectively participating in the execution of violence.[77]

Was Farid Khan, then, a 'surrogate victim', and was his lynching a public sacrifice aiming to absorb all the internal differences, doubts, and uncertainties pent up within the Naga community? Was Farid Khan's public lynching, for all its horror, a 'necessary' act – which rendered the very question of his guilt mute – through which the internal divisions, violence, and ethnic anxiety were deflected outside Naga society so that a cohesive and clear-cut sense of 'Naga-ness' could be restored, at least temporarily? If this reading of the lynching as a public sacrifice seems attractive, the abstract determinations of Girard's theory mean that it cannot be the only narrative.

For one thing, the protestors' deliberate negation of the legal and judiciary process must also be analyzed as the privatization and communalization of justice, otherwise seen as the prerogative of the state. Amid pervasive

77 Girard, *Scapegoat*.

narratives of corruption and malgovernance, the Nagaland state's inability to check illegal immigration added a particularly volatile edge to a pre-existing disenchantment, as this failure is seen as putting the very existence of the Naga ethnic category at risk. For this reason, I prefer to frame the public lynching in terms of a dramatic display of Naga ethnic sovereignty, a display that resonates with Girard's public sacrifice but also recognizes in it an assertion of popular sovereignty in the context of a failing state.

The angry mob, Lars Buur argues, turns into a 'sovereign entity par excellence' when it becomes an 'internal cleansing machine that produces temporary political communities, and divides residents into the worthy and unworthy, the trusted and the untrusted, beings and non-beings.'[78] In our case, this imperative of 'internal cleansing' manifested itself as an act of asserting Naga ethnic sovereignty. Its efficacy, however, transcended the act of violence itself, which for all its dreadfulness was 'just' a single killing. Its efficacy lay in its larger, more permanent, symbolic, and communicative powers. The visceral violence meted out on Farid Khan's body was therefore of a particular sort, a communicative act, staged in public, devoid of any ambiguity, that produced a clear notion of a unified Naga-ness in a context of increasing ethnic anxiety. By exerting the capacity to kill a 'stranger', the mob asserted Nagas' ethnic sovereignty and thus bolstered the desired ethnic order, while, in so doing, they simultaneously criticized and warned the Nagaland government about the latter's inaction.

The administering of violence on Farid Khan, as a sovereign act, did not of course permanently resolve the internal tensions and predicaments that trouble contemporary Naga society, but it nevertheless ushered in a moment of absolute clarity, and carried within itself the promise of future violence against the 'internal Other.' As a communicative and symbolic act (including its photographing, filming, and social-media distribution) its transcript was too obvious for any illegal immigrant, state official, and, lest we forget another core subtext, those Naga women who 'gamble' with 'Naga purity', to miss.

What the public lynching therefore sought to reconfigure was a set of social relations: those between Naga hosts and illegal immigrants, between citizens and the state, and between Nagas themselves – the violent removal, that is, of an 'internal Other' to reclaim the 'we.' In so doing, it sought to restore ethnic sovereignty in at least three distinct but inter-related ways: it restored ethnic sovereignty through Nagas demonstrating their capacity to degrade and dehumanize the figure of the ethnic Other on Naga soil; it

78 Buur, 'Horror', p. 28.

reinvigorated a sense of coherent 'Naga-ness' in the context of rising intra-Naga divisions and differences; and it asserted Naga popular sovereignty (of and by the people, that is) in the face of a state perceived to be incapable, perhaps even unwilling, to protect and prioritize the 'Naga good.'

Final thoughts: Mobility and ethnic tribal anxiety at the highland margins of the Indian state

The public lynching of Farid Khan offers a powerful entry into the ways ethnic anxiety can galvanize bodies and the body politic, as well as revealing the political potency of the dialectical tension between immigrant flows and popular sovereignty at the highland margins of India.[79] Majoritarianism in India, especially in the current political climate of Hindu nationalist dominance, is often associated with 'mainstream' politics and national-populism, in which Hindutva symbols, sentiments, and persuasions increasingly take over state institutions, styles of politics, and the public sphere, as well as expressing themselves through extra-constitutional acts such as vigilantism and moral policing. This usually proceeds alongside the rejection of an 'internal Other', who is perceived as a threat to the integrity, completeness, and survival of the Hindu nation, an 'empty signifier', invoking Laclau,[80] which in contemporary India has been filled with the figure of the Muslim.[81] This chapter has drawn attention to a concurrent yet differently perceived threat to majoritarianism at the physical, symbolic, and cultural margins of India in which the internal other that is rejected is the figure of the illegal immigrant and that what is at stake is tribal ethno-territorial sovereignty.

The public lynching that we discussed was a particularly macabre expression of ethnic tribal sovereignty that resulted from a particularly intolerable experience of ethnic anxiety. As such it presents an extreme case. Yet this embodied anxiety and concerns about ethnic sovereignty present

79 My analysis of these, from the perspective of Farid Khan's blood-soaked and degraded body, has not been a pretty one. What, in doing so, I may not have emphasized sufficiently is the shock and condemnation that the lynching elicited from within sections of Naga society itself (for this, see Kikon 2015b). It is these voices of condemnation and despair that show how, against many odds, a benign form of Naga identity and nationalism may still survive beneath the inchoate edges of a society long conditioned by violence and repression, beneath a nation repeatedly invaded, and beneath a splintered nationalism. Kikon, 'City of Sorrow'.

80 Laclau, *Populist Reason*.

81 See Chatterji et al., *Majoritarian State*.

an increasingly potent socio-political force in highland Northeast India. A particular source of this anxiety consists of changing demographics and political economies in what are putatively ethno-territorial homelands. For highland Asia more broadly, James C. Scott refers to this as an 'engulfment' by outsiders, often 'with state-help and capitalist financing', that now threatens 'making "hill peoples" a minority in the hills.'[82]

This contemporary dialectical interplay of neoliberal developmentalism and capitalism, the inflow of immigrant labourers, territory, and ethno-national projects has been analysed in neighbouring Myanmar and further afield in Laos, in which new (trans)national investments and capitalist projects in politically restive areas prompted the state to offer territorial concessions and with that disavowed measures of spatial sovereignty to large corporations that operate in partnership with local ethnic elites in the shape of Special Economic Zones,[83] here and there resulting in regimes of 'post-national sovereignty' (Nyiri 2012). The Indian state, in contrast, is averse to any such sovereignty concessions in a territory which remains policed as a potentially subversive zone. Instead, Northeast India witnessed neoliberalism overtaking the nature of the state itself, which came to promote a policy of 'neoliberal developmentalism' by engaging and augmenting 'the market in the North East by creating "essential conditions" like transport and connectivity infrastructure.'[84] These new neoliberal conditions and connectivities, and the mobility of labour they provoke and attract, are now imbricating into ethnic spaces, and colluding with ethnic sovereignty projects.

At times, this collusion is lethal. More often expressions of sovereignty resort to the language and force of law, public agitations, and ever volatile ethno-politics. The contemporary resurgence and politics of the Inner Line are symptomatic of the current ethnic moment. Within a month after Arunachal Pradesh was finally connected to India's vast railway network in 2015, train services had to be suspended following local fears and protests that railway connectivity would allow non-locals to bypass the necessary Inner Line regime and so threaten the local status and privileges of autochthones.[85] And just as the Indian government began investing massive amounts in a complex engineering and construction project to extend its railways into Manipur, civil society organizations in the state took to the

82 Michaud, 'What's (written) history', p. 10.

83 Woods, 'Ceasefire capitalism', p. 747-770; Rippa, 'Zomia 2.0', pp. 253-269.

84 Bhattacharya, 'Neoliberal developmentalism', p. 171.

85 Gao, 'Railways'.

streets to demand (successfully so) the implementation of an Inner Line in Manipur, and which now complicates the free flow of peoples that this railway project envisaged.[86]

In Meghalaya, meanwhile, the Khasi Hills Autonomous District Council proposed a law that would cause Khasi women to lose their right to land, property, and clan names if they marry a non-Khasi (as would the mixed progeny resulting from these unions). The purpose is 'ethnic closure', to be achieved through legislation, to prevent non-Khasis from accessing the community's land and its many lucrative mineral resources. However, influential civil society organizations object to the mining of some of these mineral resources, especially uranium, through large scale 'development-projects'. Not primarily because of the inevitable damage this brings to the environment, although this is an important subtext, but especially also because of the many non-local labourers such large-scale development projects will attract, who are seen as threatening the survival of the Khasi ethnic tribal group. Opposition to uranium mining and to several other large development projects therefore 'has little to do with development itself, or with environmental concerns, and far more to do with ethnic identity politics.'[87] These brief examples from Arunachal Pradesh, Manipur, and Meghalaya, while all unique in their local particulars, can, I argue, nevertheless be read as positioned along the same continuum of ethnic anxiety, produced by the increased mobility of surrounding lowland populations, and expressions of ethnic sovereignty of which the public lynching of Farid Khan in Nagaland represents a polar point.

In sum, for ethnic tribal communities, the new logic and language of a neoliberal form of developmentalism and capitalism that, for one thing, seeks to disassociate labour power from particular social contexts, now seems to increasingly produce the existential fear that neoliberal expansion will negate constitutionally enshrined protection on the ground and ultimately eclipse their exclusive ethno-territorial rights, identity, and belonging, and so threaten their ethnic survival. This fear has begun to stoke ethnic anxiety in unprecedented ways and now signals the emergence of an exclusionary type of ethno-politics manifested through expressions of ethnic sovereignty that reveal a particularly volatile (and always potentially violent) form of ethnic life and politics in the highland margins of India and beyond.

86 On the resurgence of the Inner-Line, see Barua, 'Return', pp. 107-120.

87 McDuie-Ra, 'Anti-development', p. 44.

References

Appadurai, Arjun. 'Death Certainty: Ethnic Uncertainty in the Era of Globalization'. *Public Culture* 10, no. 2 (1998): pp. 225–247.

Appadurai, Arjun. *The Fear of Small Numbers: An Essay on the Geography of Anger.* Durham, NC: Duke University Press, 2006.

Barua, Taz. 'Return of the Frontier: Understanding the Demands for Inner Line in Northeast India'. *Alternatives: Global, Local, Political* 42, no. 3 (2018): pp. 107–120.

Baruah, Sanjib. 'Citizens and Denizens: Ethnicity, Homelands, and the Crisis of Displacement in Northeast India'. *Journal of Refugee Studies* 16, no. 1 (2003): pp. 44–66.

Baruah, Sanjib. 'Politics of Territoriality: Indigeneity, Itinerancy and Rights in North-East India'. In *Territorial changes and territorial restructurings in the Himalayas*, edited by Joëlle Smajda, pp. 69–83. New Delhi: CNRS and Adroit Publishers, 2013.

Bhattacharya, Rakhee. 'Neoliberal Developmentalism: State Strategy in India's North-East'. In *Developmentalism as Strategy: Interrogating Post-Colonial Narratives on India's Northeast*, edited by Rakhee Bhattacharya, pp. 169–191. New Delhi: Sage, 2019.

Bora, Papori. 'Politics of Difference in the Northeast: A Feminist Reflection'. In *Northeast India: A Place of Relations*, edited by Y. Saikia and A. R. Baishya, pp. 244–261. Cambridge: Cambridge University Press, 2017.

Brass, Paul. *The Theft of an Idol: Text and Context in the Representation of Collective Violence.* Princeton, NJ: Princeton University Press, 1997.

Buur, Lars. 'The Horror of the Mob: The Violence of Imagination in South Africa'. *Critique of Anthropology* 29 (2009): pp. 27–46.

Cederlöf, Gunnel. *Founding an Empire on India's North Eastern Frontiers, 1790-1840: Climate, Commerce, Polity.* New Delhi: Oxford University Press, 2014.

Chatterji, Angana P. Thomas Blom Hansen, and Christophe Jaffrelot, eds. *Majoritarian State: How Hindu Nationalism is Changing India.* London: C. Hurst & Co., 2019.

Comaroff, Jean, and John Comaroff, eds. *Law and Disorder in the Postcolony.* Chicago, IL: University of Chicago Press, 2006.

Das, Veena, and Deborah Poole, eds. *Anthropology in the Margins of the State.* Santa Fe, NM: School of American Research Advanced Seminar Series, 2004.

'Dimapur Lynching: It was "Consensual Sex" not Rape, says Nagaland Govt Report'. *Indian Express*, 12 March 2015.

Douglas, Mary. *Purity and Danger: An Analysis of Concepts of Pollution and Taboo.* London: Routledge and Kegan Paul, 1966.

'Drive to Flush Out Migrants'. *The Telegraph*, 11 February 2015.

Dutta, Nilim. 'Dimapur Lynching: They called him a Bangladeshi and a Rapist just to Kill him' (2015), accessed 2 December 2021, https://www.dailyo.in/politics/dimapur-lynching-nagaland-sharif-uddin-khan-bangladeshi-immigrants/story/1/2543.html.

Elwin, Verrier. *Nagaland.* Shillong: The Research Department Secretariat, 1961.

Fürer-Haimendorf, Christoph, von. *The Naked Nagas.* London: Methuen and Co., 1933.

Gao, David. 'Railways and the issue of Inner Line Permit in Arunachal Pradesh: Can the two function together?'. *Economic and Political Weekly* 50, no. 8 (2015), https://www.epw.in/journal/2015/8/reports-states-web-exclusives/railways-and-issue-inner-line-permit-arunachal-pradesh, accessed 20 October 2019.

Geschiere, Peter. *The Perils of Belonging: Autochthony, Citizenship, and Exclusion in Africa and Europe.* Chicago, IL: Chicago University Press, 2009.

Girard, René. *The Scapegoat.* Baltimore, MD: Johns Hopkins University Press, 1989.

Goldstein, Daniel M. '"In our own Hands": Lynching, Justice, and the Law in Bolivia'. *American Ethnologist* 30 no. 1 (2003): pp. 22–43.

Hansen, Thomas Blom, and Finn Stepputat, eds. *Sovereign Bodies: Citizens, Migrants, and States in the Postcolonial World.* Princeton, NJ: Princeton University Press, 2005.

Haokip, Thongkholal. *India's Look East Policy and the Northeast.* New Delhi: Sage, 2015.

Heng, Geraldine, and Janadas Devan. 'State Fatherhood: The Politics of Nationalism, Sexuality, and Race in Singapore'. In *Nationalism and Sexualities*, edited by A. Parker, M. Russo, D. Sommer, and P. Yeager. New York, NY: Routledge. 1992.

Horam, Mashangthei. *The Naga Insurgency: The Last Thirty Years.* New Delhi: Cosmo Publications, 1988.

Hussain, Bonojit. 'Nation's Honour: The Dimapur Lynching and "Illegal Bangladeshi Immigrants"' (2015), accessed 2 December 2021, https://scroll.in/article/712715/nations-honour-the-dimapur-lynching-and-illegal-bangladeshi-immigrants.

Iralu, Kaka. *The Naga Saga: A Historical Account of the Sixty-Two Years Indo-Naga War and the Story of Those Who were Never Allowed to Tell it.* Kohima: K. Iralu, 2002.

Kar, Bodhisattva. 'When was the Postcolonial? A History of Policing Impossible Lines'. In *Beyond Counter-Insurgency: Breaking the Impasse in Northeast India*, edited by Sanjib Baruah, pp. 39–80. Oxford: Oxford University Press, 2009.

Karlsson, Bengt G. *Unruly Hills: A Political Ecology of India's Northeast.* New York, NY and Oxford: Berghahn, 2011.

Kikon, Dolly. 'The Predicament of Justice: 50 Years of Armed Forces Special Powers Act'. *Journal of Contemporary South Asia* 17, no. 3 (2009): pp. 271–282.

Kikon, Dolly. 'The City of Sorrow: Revisiting the 2015 Dimapur Lynching' (2015), accessed 2 December 2021, https://scroll.in/article/729595/the-city-of-sorrow-revisiting-the-2015-dimapur-lynching.

Kikon, Dolly. 'What is Unique about Naga History'. *Economic and Political Weekly* 50, no. 35 (2015): pp. 10–13.

Kikon, Dolly. *Life and Dignity: Women's Testimonies of Sexual Violence in Dimapur (Nagaland)*. Guwahati: North-Eastern Social Research Centre, 2016.

Kikon, Dolly. 'Difficult Loves: Purity and Hardship'. In *Northeast India: A Place of Relations*, edited by Y. Saikia and A. R. Biashya, pp. 225–243. Cambridge: Cambridge University Press, 2017.

Kikon, Dolly. 'From the Heart to the Plate: Debates about Dog Meat in Dimapur'. *International Institute for Asian Studies Newsletter* 77 (2017): pp. 38–39.

Laclau, Ernesto. *On Populist Reason*. London: Verso Books, 2005.

'Lawless Christian Intolerance'. *Eastern Mirror*, 9 March 2015.

Li, Tania M. 'Ethnic Cleansing, Recursive Knowledge, and the Dilemmas of Sedentarism'. *International Social Science Journal* 54, no. 173 (2002): pp. 361–371.

Maksic, Adis. *Ethnic Mobilization, Violence, and the Politics of Affect: The Serb Democratic Party and the Bosnian War*. Cham: Palgrave Macmillan, 2017.

Mamdani, Mahmood. *When Victims Become Killers: Colonialism, Nativism, and the Genocide in Rwanda*. Princeton, NJ: Princeton University Press, 2001.

McDuie-Ra, Duncan. 'Anti-Development or Identity Crisis? Misreading civil society in Meghalaya, India'. *Asian Ethnicity* 8, no. 1 (2007): pp. 43–59.

Michaud, Jean. 'What's (Written) History For? On James Scott's Zomia, Especially Chapter 6½'. *Anthropology Today* 33, no. 1 (2017): pp. 6–10.

Middleton, Townsend. 'Anxious Belongings: Anxiety and the Politics of Belonging in Subnationalist Darjeeling'. *American Anthropologist* 115, no. 4 (2013): pp. 608–621.

'Naga Council Dimapur and Naga Women Hoho Dimapur Condemn'. *Eastern Mirror*, 3 March 2015.

'Naga Mob was Wrong'. *Morung Express*, 19 March 2015.

'NSCN-IM on Jamir's Article'. *Morung Express*, 12 January 2015.

'NSF Against IBIs: Rally to be Held Today'. *Eastern Mirror*, 4 March 2015.

Nyiri, Pál. 'Enclaves of Improvement: Sovereignty and Developmentalism in the Special Zones of the China-Lao Borderlands'. *Comparative Studies in Society and History* 54, no. 3 (2012): pp. 533–562.

Ong, Aihwa. *Flexible Citizenship: The Cultural Logics of Transnationality*. Durham, NC and London: Duke University Press, 1999.

Panwar, Namrata. 'From Nationalism to Factionalism: Faultlines in the Naga Insurgency'. *Small Wars and Insurgencies* 28, no. 1 (2017): pp. 233–258.

Phizo, Angami Zapu. *Plebiscite Speech* (1951), accessed 2 December 2021,http://www.neuenhofer.de/guenter/nagaland/phizo.html.

Rippa, Alessandro. 'Zomia 2.0: Branding Remoteness and Neoliberal Connectivity in the Golden Triangle Special Economic Zone, Laos'. *Social Anthropology* 27, no. 2 (2019): pp. 253–269.

Risor, Helene. 'Twenty Hanging Dolls and a Lynching: Defacing Dangerousness and Enacting Citizenship in El Alto, Bolivia'. *Public Culture* 22, no. 3 (2010): pp. 465–485.

Roy, Sandip. 'Don't Touch "our" Women: Dimapur Lynching was never about Justice' (2015), accessed 2 December 2021, https://www.firstpost.com/india/dont-touch-our-women-the-dimapur-lynching-was-never-about-justice-2143305.html.

Simmel, George. 'The Stranger'. In Kurt Wolff (transl.), *The Sociology of Georg Simmel*, pp. 402–408. New York, NY: Free Press, 1950.

Singh, Amerjeet. *A Study on Illegal Immigration into Northeast India: The Case of Nagaland*. IDSA Occasional Paper no. 8 (2009), accessed 2 December 2021, https://idsa.in/system/files/OccasionalPaper8_NagalandIllegalImmigrationl.pdf.

Siraj, Nasreen, and Ellen Bal. '"Hunger has Brought us into this Jungle": Understanding Mobility and Immobility of Bengali Immigrants in the Chittagong Hills of Bangladeshi'. *Social Identities* 23, no. 4 (2017): pp. 396–412.

Steyn, Peter. *Zapuphizo: The Voice of the Nagas*. London: Routledge, 2002.

'Students Protest Rape in Kohima'. *The Telegraph*, 4 March 2015.

Swu, Scato. *Hails and Blames: A Brief Account of Naga Independence Struggle*. Dimapur: Heritage Publications, 2013.

Taussig, Michael. *Defacement: Public Secrecy and the Labor of the Negative*. Stanford, CA: University Press, 1999.

Tinyi, Venusa. 'The Headhunting Culture of the Nagas: Reinterpreting the Self'. In *Nagas in the 21st Century*, edited by J. J. P. Wouters and M. Heneise, pp. 117–132. Kohima and Edinburgh: The Highlander Books, 2017.

Uribe, Maria. V. 'Dismembering and Expelling: Semantics of Political Terror in Colombia'. *Public Culture* 16, no. (2004): pp. 79–96.

Van Schendel, Willem. 'The Dangers of Belonging: Tribes, Indigenous Peoples and Homelands in South Asia'. In *The Politics of Belonging in India: Becoming Adivasi*, edited by D. J. Rycroft and S. Dasgupta, pp. 19–43. London and New York, NY: Routledge, 2011.

Wood, Mary. *Lynching and Spectacle: Witnessing Racial Violence in America, 1890-1940*. Chapel Hill, NC: The University of North Carolina Press, 2009.

Woods, Kevin. 'Ceasefire Capitalism: Military-Private Partnerships, Resource Concessions and Military State Building in the Burma-China Borderlands'. *Journal of Peasant Studies* 38, no. 4 (2011): pp. 747–770.

Wouters, Jelle J. P. 'Keeping the Hill Tribes at Bay: A Critique from India's Northeast of James C. Scott's Paradigm of State Evasion'. *European Bulletin of Himalayan Research* 39 (2011): pp. 41–65.

Wouters, Jelle J. P. 'Sovereignty, Integration or Bifurcation? Troubled Histories, Contentious Territories, and the Political Horizons of the Long Lingering Naga Movement'. *Studies in History* 32, no. 1 (2016): pp. 97–116.

Wouters, Jelle J. P. *In the Shadows of Naga Insurgency: Tribes, State, and Violence in Northeast India.* New Delhi: Oxford University Press, 2018.

Wouters, Jelle J. P. 'Difficult Decolonization: Debates, Divisions, and Deaths within the Naga Uprising, 1944-1963'. *Journal of North-East India Studies* 9, no. 1 (2019): pp. 1–28.

Wouters, Jelle, J. P. 'Neoliberal Capitalism and Ethno-Territoriality in Highland Northeast India: Resource-Extraction, Capitalist Desires and Ethnic Closure'. *Geopolitics* (2020). Doi: 10.1080/14650045.2020.1812581.

Wouters, Jelle J. P., and T. B. Subba. 'The "Indian Face," India's Northeast, and "the Idea of India"'. *Asian Anthropology* 12, no. 3 (2013): pp. 126–140.

Žižek, Slavoj. *Did Somebody Say Totalitarianism? Five Interventions in the (Mis)use of a Nation.* London: Verso, 2005.

About the author

JELLE J. P. WOUTERS, Department of Social Sciences, Royal Thimphu College, Bhutan. He has conducted ethnographic and historical research in the Naga highlands and published on vernacular democracy and elections, political conflict, kinship and identity, and social history. He is author of *In the Shadows of Naga Insurgency* (2018) and has published widely in international journals.

8 Frictions and Opacities in the Myanmar-China Jade Trade

Henrik Kloppenborg Møller

Abstract
This chapter discusses how frictions and opacities are productive in the trade of jadeite jade from northern Myanmar to consumers in eastern China. The chapter first outlines a history of flows, frictions, and diversions in the Myanmar-China jade trade and opacities, taxation, and smuggling in the contemporary trade. It then presents the perspectives of a Kachin jade broker in Myanmar's Kachin State and of two jade consumers in Shanghai. The two cases illustrate how frictions and opacities structure economic strategizing and cultural imaginations. While transparency is an ideal feature of jadeite and knowledge a key asset in its trade, this chapter demonstrates how opacity and ignorance can be produced and used to serve the interests of participants in the Myanmar-China jade trade.

Keywords: jade trade, Myanmar, China, friction, opacity, ignorance

Jade refers to two distinct materials: nephrite and jadeite.[1] Jadeite jade mined in Myanmar's Kachin State has since the early 1990s seen a boom in trade volumes and prices in Mainland China.[2] This jade market boom has been influenced by several factors, including changing resource extraction

1 Jadeite is a silicate of sodium, aluminium and iron, while nephrite is a magnesium-rich amphibole. Nephrite jade has been used for at least 8,000 years in what is now China; the highest valued types mined in Xinjiang (Walker, 'Jade', pp. 22-27; Yu, *Chinese Jade*, pp. 5-7; Schumann, *Gemstones*, pp. 170-172). The vast majority of gem-grade jadeite is mined in Myanmar's Kachin State. In China, jadeite is called 'Kingfisher jade' (*feicui*) due to the similarity of its preferred emerald-green colour to that of the feathers of a kingfisher bird. See Møller, *Spectral Jade*, pp. 253-283, 330-333 for a listing and discussion of Chinese jade nomenclature.
2 The name Burma was officially changed to Myanmar in 1989. I use the contemporary name Myanmar in this chapter, except when referring to specific events before 1989.

Cederlöf, Gunnel, and Willem van Schendel (eds), *Flows and Frictions in Trans-Himalayan Spaces: Histories of Networking and Border Crossing*. Taylor & Francis Group 2022
DOI: 10.5117/9789463724371_CH08

regimes in Kachin State, intensifying Myanmar-China cross-border trade, as well as rising wealth and shifting investment patterns and consumer demand in Mainland China. While trade is often seen as predicated upon connectivity and value upon visibility, this chapter explores the flipside productivity of frictions and opacities in the Myanmar-China jade trade.

Several contributors to the anthology *War and Peace in the Borderlands of Myanmar*, edited by Mandy Sadan (2016), discuss how a ceasefire agreement between the Kachin Independence Organisation (KIO) and Myanmar's national military (*Tatmadaw*) in 1994 was followed by the latter's expansion into resource-rich areas of Kachin State, including jade mining areas, and its leasing out of resource concessions to national government and military agents, their cronies, and Chinese investors.[3] In conjunction with ecological devastation and depletion of the mines caused by an upscaling of industrialized jade mining, the cutting off of KIO and Kachin civilians from jade revenues was one factor which influenced the breakdown of the ceasefire and the resumed fighting from 2011. Control over natural resources like jade has thus played an important role in mutually constitutive cycles of war and peace, which have shaped Kachin society over time.[4] Deadly landslides, drug use, and HIV in the jade mining areas further contribute to accusations that a military-political-economic elite are the primary beneficiaries from the jade industry, while many miners and locals experience detrimental consequences.[5] Meanwhile, many of the Chinese traders and consumers, who ultimately fuel the industry, de-emphasize, silence, or are ignorant about these 'Kachin' aspects of jade biographies.

In China, jade is both an investment and a consumer object. Like other gemstones, jade constitutes a tangible, compact economic valuable, which is relatively easy to exchange and hide, and is by some investors believed to accumulate economic value more predictably than, for example, stocks.[6] These properties have made jade a popular investment object, a means for money laundering, and a currency of bribes in China.[7] As a consumer object, jade is ascribed meanings that to Chinese users make it a sort of materialisation of a resurgent 'traditional' Chinese culture that comprises a mix of Confucian, Taoist, and Buddhist philosophies.[8] In the context of a Chinese nationalism

3 Jones 'Understanding', pp. 101-107; Kiik, 'Conspiracy', pp. 217-220; Smith, 'Reflections', pp. 79, 82; Woods, 'Commercialisation'.

4 Sadan, 'Introduction', pp. 6-7, 13; Sadan, *Being and Becoming*, pp. 88-100.

5 Global Witness, *Jade*.

6 See Naylor, 'Underworld', p. 132.

7 Møller, *Spectral Jade*, pp. 191-196.

8 Ibid., pp. 37-40, 285-315.

that develops with China's global economic and political power, some users posit the popularity of jade as part of a current 'national cultural fever' (*guoxue re*). After four decades of market reforms that have also entailed opening up to foreign goods, ideas, and lifestyles, Chinese urbanites are beginning to replace Western luxury goods with what they see as traditional domestic materials and practices – including jade, tea, Kungfu, calligraphy, and Tibetan Buddhism – as sources of social distinctions and identities.[9]

Flows and material-conceptual transformations of raw jade stones excavated in Myanmar's Kachin State into luxury jewellery signifying a quintessential Chinese national identity are not always smooth and transparent. They also involve (grinding) labour, stratified mobilities, risk-taking, trust-making, hiding, withholding, and diversions. Rather than positing frictions and opacities as hampering deviations from ideal states, this chapter explores their productivity in the Myanmar-China jade trade. Anna Tsing points to how friction is productive. Involving misunderstandings, incommensurability, heterogeneous and unequal encounters, and connecting distance and difference, friction forms arrangements of culture and power.[10] In Myanmar-China borderlands, (different) local and national laws, taxation, borders, and military checkpoints define routes and procedures for legal trade and the transport of gems, but thereby also pave the way for bribery and for smuggling via alternative routes. Distance and differences in valuations are likewise productive. A jade stone that can earn a miner some weeks' worth of food in Kachin State becomes an animated bracelet and part of the identity of a woman in Shanghai. Such frictions between laws and practices, and between cultural worlds, structure earnings in the Myanmar-China jadeite trade. As Sántha, Safonova, and Sulyandziga argue about the Russia-China nephrite jade trade, companies must simultaneously bridge and keep apart two worlds to capitalize on valuation differences and hamper competition; keep jade an ordinary stone in Russia, yet turn it into a valued 'national essence' in China.[11] Cross-border jade trade requires connectivity, but distance, difference, blockages, opacities, and ignorance also constitute economic opportunities.

Robert Proctor discusses how a distinction between 'precious' and 'semiprecious' gemstones reflects modern ideals of clarity, transparency, brilliancy, purity, and homogeneity.[12] This hierarchy simultaneously segregates

9 See Osburg, 'Consuming belief'; Zhang, 'Puer Tea'.

10 Tsing, *Friction*, pp. 4-6.

11 Sántha, Safonova, and Sulyandziga, 'Searching for Trust'.

12 Proctor, 'Anti-Agate'.

opacity as an undesired quality. In gemstones, opacity denotes a lack of transparency and translucency. While high-grade jadeite is valued for its transparent and translucent vitreous texture, unprocessed jadeite stones are covered by an opaque surface crust of stone 'skin', which hides their interior jade 'meat'. This means traders cannot fully see the content of such stones before they are ground or cut open and carved. This material opacity constitutes an infrastructural indeterminacy that invites economic gambling in unprocessed jadeite stones, which are called 'gambling stones' (*dushi*) in China.[13] Opacity can broadly characterize a restriction of visibility, clarity, and knowledge. While Marxist approaches see commodity value and fetishism as hinging upon concealment, distance, and separation of productive labour from consumption, exchange theories often see value as predicated upon visibility. Combining a Marxist focus on production with Saussurean structuralism and Maussian exchange theory, Graeber thus posits value as hidden generative potentials that are realized when made visible and meaningful by being 'assigned a place in some larger system of categories' in a relatively public context.[14] What we cannot fully see and know constitutes a field of imaginative potentials that other people may profit from by contributing to, and the act of obscuring 'Kachin' aspects of jade biographies allows for construction of jade symbolism that better fits urban Chinese consumers' identity projects, and hence benefits traders. As Carey and Pedersen argue, not only is certainty an effect and dependent on an infrastructure of sorts; uncertainty and doubt must also be generated and sustained.[15]

This chapter explores imaginative and economic productivity of frictions and opacities in the Myanmar-China jade trade. The chapter first outlines flows, frictions, and diversions in the historical Myanmar-China jade trade, and taxation, smuggling, and speculation in the contemporary trade. The chapter then presents participants in two nodes of the trade: a local Jinghpaw (Kachin) jade broker in Myanmar's Kachin State, and two Chinese women in Shanghai, who animate and incorporate jadeite into their national self-perceptions, believing it originates in China. While knowledge is a key asset in the Myanmar-China jade trade, the cases show how opacity and ignorance can be produced or used to serve economic interests and the cultural self-perceptions of participants

13 Møller, 'Potentials', pp. 124-128.

14 Graeber, *Toward an Anthropological Theory*, pp. 41, 45, 70.

15 Carey and Pedersen, 'Introduction', p. 21.

in the trade. With Proctor's concept of 'agnotology', the chapter proposes to examine both the production and productivity of lack of knowledge.[16]

Flows, frictions, and diversions in the historical Burma-China jade trade

Flows of jade have connected people and places in the Burma-China borderlands and beyond for centuries but have also encountered and engendered frictions. Legend has it that jadeite jade was introduced to China in the thirteenth century by a chance discovery, when a Yunnanese muleteer in Upper Burma picked up a stone and put it in his saddlebag as a counterweight to his cargo. Back in Yunnan, the stone fell to the ground and broke, disclosing green jade inside.[17] The legend points to historical trade routes connecting Upper Burma and Yunnan, and to how the opaque surface of jadeite stones hides their valued content within.

Yang mentions that jade was traded alongside goods like ivory, horses, lumber, cloth, spices, salt, tea, gold, silver, lead, cotton, silk, and opium on caravan routes that formed part of the so-called Southwest Silk Road, which connected Yunnan with Southeast Asia, Sichuan, Tibet, Nepal, and India as early as from the second century BCE.[18] While Leach notes that Chinese works published between 350CE. and 1000CE refer to production of jade in what was likely the Kachin Hills, Sun suggests it was the invading Mongols, who, by introducing gemstones from Western Asia to China in the thirteenth century, jump-started a Chinese demand for Burmese gems that was and remains the driving force behind Myanmar's gem business.[19] But jadeite did not become widely known in China until the eighteenth century. The Chinese Ming Dynasty (1368-1644) trade with Burma was centred on rubies and sapphires (*baoshi*; gemstones). An emerging significance of jadeite jade in Qing Dynasty China (1644-1912) manifested itself in the replacement of the term *baoshi* by *feicui* ('Kingfisher jade', jadeite) in court records, which Sun dates to the year 1719. Having conquered Xinjiang, the Qing Emperor Qianlong, who reigned from 1735 to 1795, had nephrite jade mined there. Developing an obsession with jade, Qianlong wrote more than 600 poems

16 Proctor, 'Agnotology'.

17 The legend was reported by William Warry, a political officer for the British administration in Burma, previously based in Peking, in *Burma Gazetteer: Myitkyina District*.

18 Yang, *Between Wind and Clouds*, pp. 14, 43, 49.

19 Leach, *Political Systems*, p. 238; Sun 'From Baoshi to Feicui'.

about the material, which were engraved into jade, sometimes ancient pieces. Qianlong heard about an emerald-green translucent jade in Burma (which had last sent tribute to China in 1567) and waged war against its King Bodawpaya, who in 1788 dispatched envoys carrying jadeite as tribute to Qianlong.[20]

The jadeite mines were predominantly located in Jinghpaw territories of what is now Myanmar's Kachin State. Sadan discusses how an upscaling of jadeite mining and trade after the end of the Sino-Burmese wars (1765-9) engendered shifting intra-Jinghpaw economic relations and tensions and embedded these presumedly peripheral areas in world systems and events.[21] Burmese officials recognized the authority of Jinghpaw elites to tax the jade trade, which was directed eastwards towards China. The rising wealth of these 'eastern' Jinghpaw elites left Singpho elites in Assam to the west squeezed out, and some regrouped and engaged in small-scale attacks.[22] Overland jadeite trade from the Kachin hills to Yunnan intensified in following years, and William Warry, a political officer for the British administration in Burma, described the trade as prospering from 1831 to 1840. Unprocessed jadeite stones were sent to an emporium in Yunnanfu (Kunming), where they were bought especially by Cantonese merchants, who sent them to Canton (Guangdong) to be cut and carved. The start of the first Opium War in 1839 stopped Cantonese merchants from going to Kunming to buy jade and therefore also Yunnanese traders from going to the mines in Kachin State. This led to a crisis for Jinghpaw mine owners and miners, and Sadan suggests a connection between the Opium Wars in Canton and a widespread Singpho-Jinghpaw revolt against the British East Asia Company in Assam in 1843.[23] In China, the Taiping revolt from 1850 and the Panthay revolt from 1856 again disrupted overland trade routes. William Warry reported that a Cantonese merchant bought up the stocks of jade in Mandalay in 1861 and made a fortune by shipping the jade to Canton from lower coastal Burma. The shipping route revived the trade. This underlines how traders could profit from diverting flows of jade in response to disruptive events like wars and how ethnic Chinese in Myanmar have served as middlemen in the trade. Cantonese are still strongly involved in jadeite trading, and the largest current Mainland Chinese jadeite markets

20 Levy and Scott-Clark, *Stone of Heaven*, p. 39.

21 Sadan, *Being and Becoming*, pp. 88-100.

22 The Jinghpaw (Kachin) in Myanmar, the Singpho in Northeast India, and the Jingpo in China's Yunnan Province are considered parts of the same main ethnic group.

23 Sadan, *Being and Becoming*, pp. 90-91.

outside Yunnan are in Guangzhou, Jieyang, Sihui, and Pingzhou in China's Guangdong Province.

Like Qianlong, the Dowager Empress Cixi, who effectively controlled the Chinese Qing government from 1861 until her death in 1908, developed a strong appetite for jade focusing on its perceived aesthetic, spiritual, and medicinal properties. Taoist scholars drank elixirs containing crushed jade in an attempt to attain immortality and wisdom and stimulate sexual potency, and Cixi instructed her eunuchs 'to collect dew in a bronze plate that was mixed with powdered jade into an elixir of life'.[24] By 1875, Cixi had a steady supply of jadeite, and she commissioned a treasure hall to store it. But people from the court stole jade pieces, and shops outside the Tiananmen gate in Peking exported cases of jade to London, Paris, and New York, feeding an international market kickstarted by the Allied plunder of the Forbidden City in 1860 after the second Opium War.[25] In the 1920s and 1930s, jade jewellery had become highly treasured among socialites in Shanghai, London, and Paris,[26] and, though incomparable to Chinese demand, jade remains in demand in the West as investment, art, jewellery, and as a healing material among subscribers to New Age philosophies.[27]

Chiang Kai-shek, leader of China's nationalist Kuomintang (KMT) Party, had looted Cixi's citadel at the Imperial necropolis and Peking's Imperial Warehouse for jade, which he moved to Taiwan as the KMT lost the Chinese civil war to the communists. In 1950, KMT forces isolated in Yunnan crossed the Burma border and established bases in Shan State. Supplied with weapons by the US Central Intelligence Agency (CIA), they struck China's communist troops. The Kachin Independence Army (KIA), established in 1961, formed a jade company and bought such CIA-supplied weapons from the KMT in exchange for jade to combat the Burmese army for greater autonomy.[28] During Mao's communist reign (1949-1976), demand for jade withered in Mainland China, where it was associated with bourgeois decadence, especially during the Cultural Revolution (1966-1976). Cross-border trade was formally restricted in Burma's socialist period (1962-1988), and much jade was smuggled via the Burma's Shan State to markets in northern Thailand, which catered especially to buyers from Taiwan and Hong Kong. Yunnanese caravan traders in Burma employed borderland knowledge,

24 Levy and Scott-Clark, *Stone of Heaven*, pp. 93, 159.

25 Ibid., p. 156.

26 Ibid., pp. 224, 237.

27 Møller, *Spectral Jade*, pp. 50-51.

28 Levy and Scott-Clark, *Stone of Heaven*, p. 284.

networks, mobility, mercantile agency, and risk-taking in conducting such illegal cross-border trade in interaction with political-military entities like KMT, Shan, Wa, and Kachin troops.[29] Among others, Khun Sa, a Chinese Shan, who headed the Shan United Army and operated a drug cartel from the mid-1970s to the mid-1990s that allegedly at one point supplied a quarter of the world's heroin, offered protection to caravans from Shan State to Thailand in exchange for jadeite.[30] Restrictive nation-state ideologies and trade policies in China and Burma thus paved the way for a risky but profitable underground jade trade for those with the means to navigate the politically and militarily complex borderlands of northern Burma.

In China, the first decade of Deng Xiaoping's economic reforms from 1978 unleashed explosive entrepreneurial activity and economic growth in eastern coastal Chinese cities and provinces, which created the basis for a renewed Chinese consumer demand for luxury goods like jade. But it was only in the early 1990s that 'promises of mobility, trade, and investment embedded in Deng Xiaoping's reforms took concrete shape' in western inland Yunnan Province.[31] As China eased restrictions on border trade, and demand shifted from Hong Kong and Taiwan towards Mainland China, the importance of Yangon and Chiang Mai as transit stations decreased, and major jadeite flows were diverted back to overland routes in the Myanmar-Yunnan borderlands.[32] The border town of Ruili in Yunnan, neighbouring Myanmar's Shan State, became the main official entry port for jadeite imports to China.

Traders in Ruili say that jade sales volumes and prices boomed from 2009, and in June 2016, the Myanmar Times reported that 300,000 miners were working at the jade mines in Kachin State. This jade boom has occurred during US and EU sanctions that targeted Myanmar's mining, timber, and gemstone industries in reaction to the government's violent suppression of the 2007 Saffron Revolution, intending to emulate Africa's 'blood diamonds' model.[33] The jade boom underlines the continued leading part of Chinese demand in Myanmar's jade industry. Currently, Yunnan holds strategic importance for China's USD 1 trillion Belt and Road Initiative (BRI) and hosts several BRI projects, including the China-Myanmar Economic Corridor (CMEC), which comprises infrastructure construction, trade deals, and

29 Chang, 'From a Shiji Episode'; Chang, *Beyond Borders*, pp. 149-175.

30 Levy and Scott-Clark, *Stone of Heaven*, p. 284.

31 Rippa, *Borderland Infrastructures*, p. 23.

32 Frey and Lindtner, 'Jade', p. 271.

33 Egreteau, 'Burmese Jade Trail'.

special economic zones to upscale China's access to markets and resources in Myanmar and beyond.

This historical overview should illustrate how the Burma-China jadeite trade has developed along ancient regional trade routes and, as Sadan discusses, in interactions with global events and emerging world systems of economic and imperial power.[34] Jadeite has been mined in Kachin lands and known in Yunnan for many centuries but became globalized after being exported to Peking as imperial tribute in the late eighteenth century. Its transparency, translucency, vitreous texture, and emerald-green colour distinguishes 'Imperial-grade' jadeite from the opaquer, more lacklustre, and coarser nephrite. But by conceptually subsuming jadeite as a sub-type of jade (*yu*), Chinese users ascribed the same spiritual and medicinal properties to jadeite as to the domestically mined nephrite. Chinese demand upscaled mining and trade, and thereby also economic stratification and conflicts, and jadeite has been a protagonist in wars; from Qianlong's war against Burma to its use by the KIA as payment for weapons from the KMT and recent clashes between the KIA and the Tatmadaw for control of the mines. Tina Harris uses the term 'geographical diversions' to highlight processes whereby some places and routes are brought into view while others are obscured, which, in turn, creates diversions to other places and routes.[35] Frictions and disruptions due to events such as wars have engendered diversions of jadeite trade routes. But the closing of borders and diversion to sea routes has also offered opportunities for small-scale trade and smuggling along overland routes in Myanmar-Yunnan borderlands that are still used to evade taxation today.

Opacity, taxation, and smuggling in the contemporary Myanmar-China jade trade

Woods coined the term 'ceasefire capitalism' to address how Myanmar's government used the 1994 ceasefire agreement with the KIO to gain control over and sell off natural resources in Kachin State.[36] Since the mid-2000s, the only legal way for foreigners to buy and export unprocessed jade is through emporiums organized by Myanmar Gem Enterprises, a state-owned enterprise that regulates the gemstones sector and trades gemstones

34 Sadan, *Being and Becoming*.
35 Harris, *Geographical Diversions*.
36 Woods, 'Ceasefire Capitalism'.

through joint ventures with private companies. The government ban on exporting unprocessed jade not bought at the emporiums marks an attempt to monopolize mining licensing, trade, and taxation. This tactic mirrors that applied to the timber industry, which was previously in the hands of Kachin groups, but has since 2006 been appropriated by the state-owned Myanmar Timber Enterprise. This entailed that timber exported to China without government permission is defined as 'illegal', and that 'legal' timber became a 'national' resource largely out-of-bounds to the KIO and local Kachin people.[37]

A Chinese government official in Ruili told me that control over the jade mines was a primary factor in the breakdown of the 1994 ceasefire agreement in Kachin State in 2011. He traded jade in 1992-1993, when the KIA controlled most mines. He said that the government's takeover of mines intensified from 2008, and that jade prices in Ruili boomed from 2009 to 2013. Resumed fighting from 2011 saw the Tatmadaw conquering new jade mines from the KIA, but it also halted mining, and official jade production dropped from 43.2 million kilograms in 2011 to 15.1 million kilograms in 2013. In 2014, the Myanmar government allowed companies to resume mining, and jade production rose to 37.7 million kilograms in 2015.[38] Having been paused due to the fighting, the government gem emporium was held in Mandalay in mid-June 2013. An entrance ticket demands a EUR 50,000 deposit, and the attendees at the emporium were mainly Chinese top-level traders. Posts on the mobile phone app WeChat informed traders staying behind in Ruili about supply and prices. Two traders in Ruili said that jade prices had increased by 30 per cent from February to May 2013, but that trade slowed down in June, as traders waited to see sales prices at the emporium. If prices were very high in Mandalay, traders would hold on to their best stones in anticipation that prices would further increase. If prices were low, they would sell out of their stocks.

Jade flows can be divided into southern and northern routes, of which the latter are used most today. The southern route sees jade transported southwards to Yangon and shipped out, or via Shan State to northern Thailand. One branch of the northern overland route sees jade transported to Mandalay and then northeast via Shan State to Ruili in China. KIA-controlled stones are usually smuggled into Yunnan via routes that cross the border at, or near, Mai Jia Yang, Pianma, Pangwa, Kambaiti, Tengchong, Laiza,

37 Woods, 'Commercialisation', p. 138.

38 These figures are from the Myanmar Statistical Information Service, as cited by Openjadedata.org (http://www.openjadedata.org/Stories/how_much_jade_worth.html).

Map 8.1 Jade flows across Myanmar and China

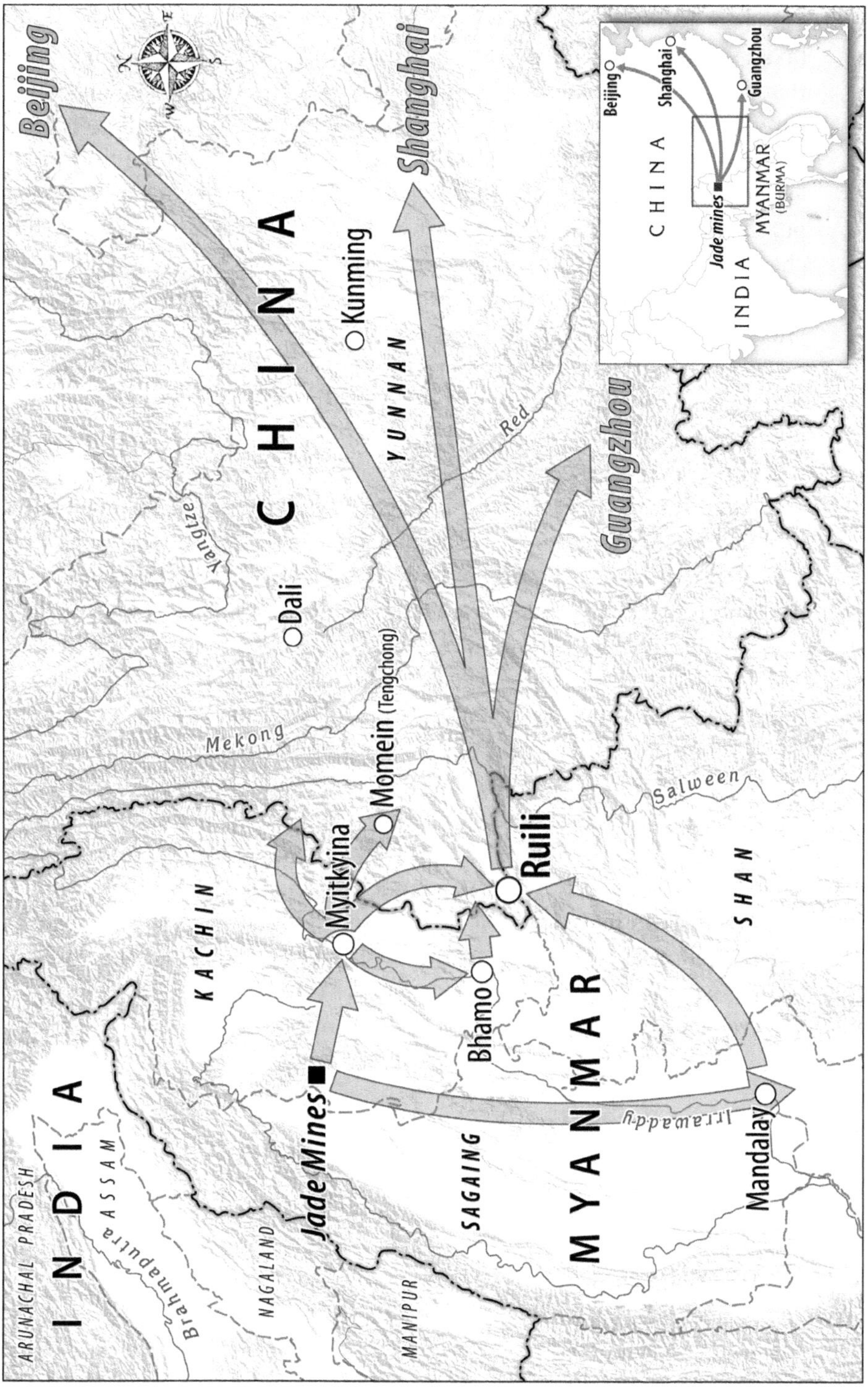

Drawn by Laurie Whiddon

Yingjiang, Nongdao, and Ruili. A jade trader in Ruili distinguished between taxation on southern and northern routes thus: 'Traders who buy jade stones that go via Mandalay and Yangon to Hong Kong, Taiwan, and Thailand pay taxes to the Myanmar government. Traders who buy stones that cross the Myanmar-Yunnan border pay taxes to the KIA. Many traders in Ruili have long-term relations with the Kachin people'. Taxes can be included in the purchase price for KIA-controlled stones but may also be levied at a KIA checkpoint before the Chinese border. While defined as illegal by Myanmar government laws, stones that are not taxed or only taxed by the KIA are effectively legalized in China if subjected to import tax.

Traders in Ruili say that buyers of government-controlled stones in Myanmar must pay a total of 33 per cent of their estimated value in government tax for them to be transported legally to China, while 34 per cent of sales value import tax is levied at the Chinese border.[39] A report by Global Witness estimated the value of jade production in Myanmar in 2014 to have been as much as USD 31 billion, equating to 48 per cent of Myanmar's official GDP that year.[40] Based on figures for jade production and emporium sales prices in 2014, Global Witness researchers calculated a discrepancy of up to USD 6 billion between actual government tax income and what taxes could have been generated that year.[41] Jade companies often bribe officials to reduce or avoid taxes levied at the mines.[42] A means of lowering taxes at the emporiums is to grind open a 'window' into the opaque surface crust of stone at a position that exposes a poor-quality part of its jadeite content, which sets the stone's taxable value at a fraction of its real value. The company's agents or proxies then secure the winning bid. Because the sales price is deflated, the tax paid is far lower than it should be. But having been taxed, the stone can be legally exported to China, where the company opens windows in the stone that display its most valuable parts and sells it at a much higher price. An industry source estimated to Global Witness researchers that around 60 per cent of the 2014 emporium sales involved

39 In 1934, H. L. Chibber likewise reported that jade stones could only leave Kachin State after payment of an *ad valorem* royalty of 33 per cent to the Government Jade License (cited in Hughes 1999). Meanwhile Global Witness, *Jade*, pp. 34-35, reports total taxes in Myanmar to be 30 per cent, divided into 20 per cent of estimated value and 10 per cent of sales value.

40 Global Witness, *Jade*, p. 6

41 Ibid., p. 35.

42 William Warry reported on under-valuing by official jade appraisers back in 1888: 'The value of jade was determined for purposes of taxation by an official appraiser. This officer, however, by private arrangement with the traders and the Collector, estimated all stone about one-third of its real value' (cited in Hughes, *Burma's Jade Mines*).

vendors buying their own jade and that top companies use this method for around 80 per cent of the jade they sell.[43] The material opacity of the surface crust covering jadeite stones can thus be productive in facilitating a reduction in taxes. Experienced traders employ knowledge about surface-substance relations of stones from different mines and geological layers to infer content of jadeite stones by examining their surface crust, but the interior content of stones is only fully revealed when they are cut and carved in China.[44]

Some buyers who have stockpiled a certain type of jade place high bids on similar stones at the emporiums, but do not pay and collect their winning bids. This enables them to restrict supply and drive up prices for their own stones. Big companies mostly sell low-grade jade at the emporiums, alongside a token amount of good quality jade 'for show'. In turn, top-grade jadeite stones are usually smuggled, and Global Witness researchers estimated that between 50 and 80 per cent of jade bypasses the emporiums and is smuggled untaxed to China.[45]

Control over the jadeite mining, trade, and taxation remains a source of conflict in Kachin State today. The friction of armed conflict has at times halted mining, but restriction of supply may also drive up prices.[46] While both the Myanmar government and the KIA tax jade mining and trade, traders engage in smuggling and bribing to cut expenses, conduct speculative bidding to drive up prices, and withhold and sell jade at different times depending on price development. The opacity of the surface crust of jadeite stones is productive in allowing traders to reduce tax expenses by grinding 'windows' into positions in the surface crust of stones that expose low-quality jadeite.

Below, I present a Kachin jade broker's articulation of frictions deriving from Chinese involvement in natural resource extraction in Myanmar's Kachin State, as well as his strategizing on a presumed lack of knowledge among Chinese amber traders.

43 Global Witness, Jade, p. 35.

44 In 1934, H.L. Chibber reported that in the Kachin mining areas 'the stones are not cut until after the Government royalty has been paid on them, and therefore their real value is merely guess-work' (cited in Hughes, *Burma's Jade Mines*). Naylor, 'Underworld', p. 146, similarly notes that Thai firms 'export ruby while still in the rock matrix – that [...] makes it easier for mine owners to undervalue the shipment for export-tax purposes'.

45 Global Witness, *Jade*, p. 24.

46 Møller, *Spectral Jade*, p. 195.

The knockoff amber machine: Imagining gem trade in Myanmar's Kachin state

I am drinking Burmese milk coffee in a teahouse overlooking the river Ayeyarwady, when a young man sporting black Ray-Ban sunglasses parks a loudly crackling 250 cc Honda dirt bike, and addresses me: 'Hey man, welcome to Myitkyina!' Jack is a local ethnic Jinghpaw (Kachin).[47] He holds a bachelor's degree in English and has worked for a foreign NGO but lost this job recently. I soon come to appreciate Jack's English skills, personality, and local knowledge, and for the next three weeks, we cruise Myitkyina's dusty roads on the Honda dirt bike, stopping intermittently at teashops, markets, and private homes to interview gem traders. On the motorbike, Jack comments on our environment: 'See those two fine Kachin women straight ahead? They will probably marry Chinese men because China has too few women'. When we pass logging trucks or pyramid-shaped gravel mounds indicating illegal gold dredging, Jack mutters 'why do they cut down all our trees here, don't they have trees in China?' or 'our land is so rich, but the Chinese [expletive] take everything'

The 1994-2011 ceasefire between the KIO and the Burmese military was characterized by unbridled logging, gold dredging, and gemstone mining in Kachin State, and much of the business was controlled by the government and Chinese interests.[48] Wouters' chapter in this volume shows how for the Naga of northeast India, the inflow of immigrants plays into an ethnic nationalism that holds women's reproductive bodies as sacred for upholding the ethnic nation's purity and honour, while state construction of Rohingya as outsiders in Myanmar discussed in Than and May's chapter is supported by Buddhist Burman nationalism. Jack voiced an injustice regarding appropriation of both Kachin natural resources and Kachin wives by Chinese businessmen, who hold greater economic power and political connections than the Bengali immigrants and Arakanese Muslims in Wouters' and Than and May's cases. Myanmar-Yunnan borderlands may be considered 'contact zones', defined as 'social spaces where cultures meet, clash, and grapple with each other, often in contexts of highly asymmetrical relations of power'.[49] China's Belt and Road Initiative intensifies contact through infrastructure construction, but while roads are often celebrated

47 Classifications of Kachin sub-groups vary. According to Jack, there are six major sub-groups: Jinghpaw, Zaiwa, Maru/Lawngwaw, Lisu, Lashi/Lachik, and Rawang.

48 Smith, 'Reflections', p. 79; Woods, 'Commercialisation', pp. 133-134; Kiik, 'Conspiracy', p. 218.

49 Pratt, 'Arts', p. 34.

as harbingers of connectivity, they may also increase social distances.[50] Just as modern natural science has denigrated opacity in favor of visibility and clarity, globalization proponents risk abandoning friction, social distance, and mistrust as normative opposites to connectivity, intimacy, and trust, rather than exploring their productivity as social phenomena. A volume on China-Russia borderlands edited by Humphrey thus explores how mistrust can be socially productive, for example by allowing mediators to emerge.[51]

Risks and mistrust in the Myanmar-China gem trades make knowledge and trust resources that configure economic opportunities. Jack occasionally earns money as a *bwesa* (broker) in the gem trade. This position requires trust, but not necessarily in-depth gemstone knowledge. Jack explains:

> I have a high reputation here. I don't know about jade and amber, but people can trust me. I have some friends who do jade or amber as a side business. When they are busy, they ask if I can sell this [gemstone]. I get a percentage of the sales price. If my friend wants 1 lak [100,000 Kyat; 66 USD] and I can sell it for 1.5 lak, I can keep the 0.5 lak.[52] That's brokerage. It depends on your clever and smooth talk, making impression on buyers. The broker doesn't need much effort, what you invest is your talk. You use smooth talk, so people can tune their minds into the story you tell. Like, 'this jade stone is cheap, you can turn it into a bracelet and make money'. We have so many marketing techniques.

We interview bwesa, who meet sellers and buyers in teahouses. Some jade stones pass through 4-5 bwesa. If a buyer bids on a stone, the first bwesa calls the owner of the stone, who decides about the price. While the bwesa's key skill is to make buyers believe in, or 'see', economic potential in particular jades, determining their authenticity and quality depends on the buyer's jade knowledge, which in turn affects his reputation. Jack tells a story that illustrates this principle:

> There is a famous story told by jade traders. Eight or nine years ago [in 2006-2007], a Chinese guy brought a prestigious jade to Hpakant [the mining area]. He gave it to his local bwesa to sell for him. The bwesa showed it to everybody. They thought it was from a local jade mine. Finally, a local

50 Pedersen and Bunkenborg, 'Roads'.

51 Humphrey, 'Introduction'.

52 The exchange rate was approximately 1 USD to 1,500 Myanmar Kyat during my research in Myanmar.

> rich guy bought it for millions of dollars and brought it to China to resell. He found out it was a knockoff! He lost everything, went down overnight. He lost his prestige and had to start his jade life again. The bwesa told him, 'it's not my fault. You are the buyer, you should carefully study [the jade] before you buy'. If you are cheated, it means your knowledge is not good enough, it is very shameful in the jade community. Fame is important. If you have fame, all jade buyers and sellers trust you. People do business with people they trust. Knockoff jade from China started spreading here after that Chinese introduced it. Some say they use low-quality jade to make knockoffs [by treating it with chemicals and colour].

Consumer perceptions of the authenticity of many types of goods presuppose that they are found, cultivated, or produced on particular sites.[53] Jack says that counterfeit jade was unknown in Kachin State before 2006, and selling a counterfeited jade in the mining area thus made it more probable that buyers would consider it genuine than in Chinese markets, where counterfeit jade is widespread. It is up to the buyer to determine the authenticity and quality of a jade. For the buyer, knowledge therefore trumps trust in people. Meanwhile, trust in people is essential for the seller, who gives his jade to brokers to sell and to carriers to transport. As mediators between sellers and buyers, brokers and carriers mitigate risks and mistrust in the trade.

Jack and I interview Kachin traders at a jade market in Myitkyina. One trader estimates that the government now (in 2015) controls 75 per cent of mines and the KIA 25 per cent. Traders who are members of a jade trade association, buy jade from government-controlled mines, and pay taxes can apply for a certificate from the Ministry of Mining that allows them to export the jade. Traders without export permission usually use 'carriers' to smuggle their jade, the carriers' salaries varying with the amount and quality of the jade smuggled. After giving jade to the carrier, the trader goes to China and waits for the carrier. This arrangement reduces the risk of imprisonment for the trader but requires trust in the carrier. The carriers have cultivated relationships with the KIA, Burmese military and police, and Chinese immigration officers, and bribe their way through. First, the carrier goes to the checkpoint without the jade, negotiates the bribe with the officers and pays it in advance. Later, he brings the jade by car or motorbike, and the officer will not 'find' it. But carriers without the right connections who are caught smuggling by Burmese military or police risk being imprisoned. Jade flows encounter frictions due to national laws, taxation regimes, and

53 Miller and Davis, *Intellectual Property*; Vann, 'Production Matters'.

borders, but relations between gem traders and smugglers, and military, police, border guards, and officials complicate rigid oppositions between state and non-state and between licit and illicit.

Another Kachin trader tells us that locals have recently found jade in the far north of Kachin State. There is no industrial machinery there, but locals may find good stones even with basic tools. He says these stones are 'outside politics', which means that neither the Tatmadaw nor the KIA controls or taxes mining there yet. After fighting re-erupted in 2011, some locals started 'digging in the shadow': 'When the political situation is fixed, jade business is difficult', the trader says, 'but when there is fighting poor people can dig, because there is no control'. Some of his friends mine jade. A boss pays for their lodging and food, and they split profits from stones fifty-fifty with the boss. He says there are more than ten checkpoints on the road to the Chinese border at Kambaiti. Burmese officers can be bribed, but Chinese border officials are harder to bribe, so his friends smuggle the stones across the border on small paths by motorbike. The government also controls the road from Myitkyina to Laiza, where the KIA controls an area around the Chinese border. Border crossing is difficult here, because there is fighting, and security is tighter. Some stones are smuggled via Bhamo to Muse in Shan State and then into Ruili in China. The trader describes his first visit to Ruili: 'I was shocked! There is so much jade there. All is from our Kachin State. It was sad to see'. Digging up a jade stone, smuggling it, selling it, carving it, or wearing a piece of jade jewellery comprises restricted fields of vision – not seeing the forest for the trees, so to speak. Crossing the national border to Ruili to witness the extent of Chinese imports of Kachin natural resources constituted a shocking widening of vision for this trader.

Due to a national anti-corruption campaign and falling economic growth in China, demand for high-grade jadeite declined from around 2014, followed by exploding demand for amber: a cheaper material also mined in Kachin State.[54] This engendered frantic amber mining, especially by ethnic Lisu, outside Tanai northwest of Myitkyina. The KIA controls most amber mines, but some miners just go to the forest and dig. Jack and I interview a Lisu herbalist, who says burning amber and inhaling the smoke can cure ailments. He leases an amber 'hole' (mining pit) and sells the amber in Myitkyina and in Tachileik on the Shan State border with Thailand. Another Lisu, who leases five amber holes, says Chinese businessmen rent mining holes and provide equipment, but hire Lisu workers to do the mining. Most holes are 8-10 metres deep, but some as deep as 50 metres. From three metres

54 Møller, *Spectral Jade*, pp. 159-161, 198-203.

down, oxygen is provided through a generator-powered air tube. This mine owner pays a carrier to bring the amber to Myitkyina. The KIA and the Tatmadaw control different sides of the entrance to the amber mines, and both levy taxes. The carrier's fee depends on the size, weight, and quality of the amber and the risks involved. Similar to jade transport, taxes can be reduced by hiding expensive amber pieces among cheap ones. He sells his best amber in Tengchong in Yunnan, where amber prices are around 25 per cent higher than in Myitkyina. Recently, the amber mining area has seen fighting between the KIA and the Tatmadaw, and the report *Blood Amber* indicates that the amber industry is undergoing a trajectory similar to that of the jade industry.[55] As Chinese demand soars, armed clashes intensify, the Tatmadaw conquers mines and sells mining licenses to outsiders, and locals risk losing their land.

Our examination of jade and amber markets in Myitkyina is coming to an end, when Jack presents the following plan to me:

> Now amber is the hottest commodity. Demand is still higher than supply, so now is a good time! But the market is already established. People have mining holes, and the Chinese control the market in China. It is a difficult market to enter. So how do we do it? We sell knock-off amber to the Chinese!
>
> The Chinese at the mines don't know much about amber, they can easily be cheated. So you must import a machine to Myitkyina that can make fake amber. The Chinese introduced fake jade here eight years ago, so people are careful about jade. But until now, nobody heard about fake amber, so it's very important to keep that word silent. Don't say 'fake' and 'amber' in the same sentence, because then people will become suspicious. Maybe there is fake amber in China, but not here. You must find a knock-off specialist in Denmark, so I can study from him.

Proliferation of new people and goods in a society can create anxieties about who and what is considered genuine.[56] Jack says that Chinese traders profited from a lack of knowledge about counterfeit jade in Kachin State 8-9 years ago, and now he proposes to similarly exploit a presumed lack of knowledge about counterfeit amber among Chinese traders here. Both knowledge and lack of knowledge are exploitable in the Myanmar-China gem trade, and Jack adds ingenuity to market knowledge – another of his proposals is to train

55 Kachin Development Network Group (KDNG), *Blood Amber*.
56 Notar, 'Authenticity Anxiety'.

a monkey to find amber in the forest. But Jack operates in the margins of a market structured by Myanmar's government and military, big companies, and Chinese consumer demand. Changing militarized control of jade and amber mines and upscaling from artisanal to industrial mining favours Burmese and Chinese companies with investment capital and connections, while many locals are directed to work as small-scale traders, smugglers, and middlemen, or 'digging in the shadow'. Meanwhile, Burmese Chinese in Myanmar employ language skills, connections, economic capital, and high mobility to link jade suppliers with buyers in China.[57]

The investment capital, connections, mobility, and thus also field of vision of small-scale Kachin gem traders like Jack are more restricted. Jack has never been to China, does not have connections there, and does not speak Chinese. But rather than neglecting the opacity of gem commodity chains that Jack faces as a normative anomaly to an ideal of perfect market visibility we may see his knockoff-amber proposal as an example of how opacity can be productive for both cultural imaginations about ethnic others and for economic strategizing in Myanmar-China borderlands. Below, I discuss how opacity is similarly productive for cultural imaginations among Chinese jade users.

National identity and ignorance among jade users in Shanghai

Chinese demand feeds on narratives that animate jade and support consumers' images of themselves and their social worlds. Such narratives include some information and exclude other information. Knowledge about jade is significant for consumers' self-image, but so is ignorance.

I interview Cindy and Chloe, both in their early 30s, in a café in Shanghai's French Concession. A native Shanghainese, Cindy holds a MA degree in English and works for a British market research company. Chloe is from northern China, has a MA degree in sociology from a British university and works in marketing. Both wear jade jewellery. Cindy expresses a need for knowledge about products she wears to avoid being seen as a fraud:

> If you don't have knowledge about what you are wearing, it becomes similar to fake stuff. I don't wear jade Buddha jewellery, because I don't know enough about Buddhism yet. It is like Louis Vuitton customers: they feel they have to know about the brand before wearing it.

57 Møller, *Spectral Jade*, pp. 143-146.

Acquiring knowledge about foreign luxury brands like Louis Vuitton is a way for Chinese consumers to build confidence that others will consider products they use, and thereby their self-signification as consumers, to be genuine in a market saturated with counterfeits.[58] Many Chinese women receive, and learn about, jade from their mothers. Jade knowledge is also communicated by vendors, advertising, and popular and scientific dissemination. Coupling jade with Chinese civilisation as iterated in the current nation-state, this information is partial and normative. Akin to how Sigley describes a bourgeoning Chinese 'tea nationalism', current signification of jade as a symbol of Chinese nationhood feeds a Chinese 'jade nationalism'.[59] Chinese jade scholar Ming Yu thus describes jade in patriotic fashion: 'Its history constitutes an epic that records the artistic creativity and indomitable spirit of the Chinese nation and inspires the Chinese people of today to strive increasingly for greatness'.[60] Chloe says that jade expresses her 'Chinese identity' and that she wears jade jewellery abroad to 'distinguish' herself.

As heirlooms and gifts, jade is also seen to host biographies of and obligations towards family members.[61] Cindy wears a jade bracelet that belonged to her great-grandmother, also when she sleeps. She feels she cannot take it off because her deceased great-grandmother would feel it. 'I feel I have an obligation to wear it. I have to nourish it by wearing it. When I get older the jade will nourish and protect me.' For both women, wearing jade also marks a turn to adulthood that coincides with their reconnection with traditional Chinese culture. Chloe compares jade with the old poems they had to recite, and the calligraphy they had to write in school. 'This is traditional Chinese culture. You cannot appreciate it when you are a child, but when you grow up you start to appreciate it.' She posits a reconstruction of jade as a token of a traditional, authentic Chinese culture in recent years as a response to rapid societal changes in China:

> The revival of jade happened at the same time as the revival of Buddhism. People are tired of things changing too fast. We are rooted in Confucianism; we want things to progress in a natural way. For 100 years, we put Western culture over our own culture. Now China's power is rising, and Chinese people are picking up our old culture. Many people start to play traditional

58 Møller, *Becoming*, pp. 79-80.

59 Sigley, 'Tea'.

60 Yu, *Chinese Jade*, p. 144.

61 Møller, 'Potentials', p. 133.

> Chinese music instruments, do calligraphy, going to teahouses. There is a 'national fever' [guoxue re] now. It started around five years ago [in 2010]. I also started to appreciate traditional Chinese culture around this time.

These narratives about how jade symbolizes and enforces family relations, national culture, and identity in a rapidly changing society echo those of other Chinese respondents I interviewed in Shanghai. I ask Cindy and Chloe whether they know where jadeite jade is mined. Also echoing other respondents in Shanghai, they say jadeite is mined in China.

Why and how does some information about jade become prevalent in China, while other information does not? Robert Proctor coined the term 'agnotology' as the study of ignorance making. Ignorance as a 'native state' implies a kind of 'original deficit' not yet penetrated by knowledge, which spurs education and science to remedy it.[62] Views of ignorance as a 'selective choice' recognize that focusing on some issues implies ignoring others, and 'once things are made unknown – by suppression or by apathy – they can often remain unknown without further effort'.[63] Finally, ignorance as 'strategic ploy' is actively and often deliberately produced, as in the tobacco industry's efforts to manufacture doubt about the hazards of smoking. I found that Chinese jade vendors selectively highlighted history, culture, and symbolism of jade in China, while omitting information about jade mining and trade in Myanmar's Kachin State from their narratives.

While diamond company DeBeers made diamonds 'a girl's best friend' and a symbol of marriage in the West,[64] jade has been re-constructed in post-Mao China as a symbol of a quintessential 'Chinese identity' that extends almost seamlessly from individual via family to nation-state. Purifying such symbolic connotations of gemstones involves filtering out 'dirty' aspects of their biographies, and the Kimberley process's work to eradicate African 'blood diamonds' made DeBeers prosper by presenting a supposedly 'clean' alternative.[65] Woods traces connections between timber production in Kachin State, processing along the China-Myanmar border, and consumption in eastern China, revealing how the violence of the insurgent economy in Myanmar 'becomes erased as the commodity becomes globalized as a product to be consumed'.[66] Walsh discusses how

62 Proctor, 'Agnotology', p. 5.
63 Ibid., p. 8.
64 Proctor, 'Anti-Agate', pp. 390-391.
65 Naylor, 'Underworld'.
66 Woods, 'Conflict Timber', p. 502.

Malagasy sapphire miners and traders speculate about the uses to which foreigners put the sapphires, and points to 'knowledge differentials' between Malagasy miners and foreign traders.[67] In these cases, aspects of diamond, timber, and sapphire biographies are hidden from the view of different actors. This is also the case for jadeite. Combined with the subsumption of jadeite as a sub-type of jade (*yu*), the erasure, downplaying, or tainting of its 'Kachin' biographies allows Chinese buyers to value jadeite as a symbol of Chinese identity, even though it is not mined in China.

Production and productivity of frictions and opacities

Introducing this volume, Van Schendel and Cederlöf stress the need for understanding how flows and frictions in the India-China Corridor are dynamically intertwined and produce social arrangements over time. While conditioned by state powers, commodity flows spanning northeast India, Burma, and Yunnan have historically unfolded along ancient trade routes and through local and regional networks and intermediaries, as the contributions by Pachuau, Ma, and Cederlöf show.

Fields of vision and inquiry highlight some phenomena and shade others. Normative visions of globalisation and free trade may accentuate flows and connectivity over frictions, while modern ideals of transparency and visibility have segregated opacity as an undesired quality in gemstones. While trade is often seen as predicated on flows and value on visibility, this chapter has explored how frictions and opacities are produced and productive in the Myanmar-China jade trade. The chapter has sought to demonstrate how frictions in the form of wars, borders, taxation, restrictive trade policies, and mistrust do not simply block jade flows, but engender diversions to alternative routes, smuggling, bribery, and middlemen positions such as jade brokers and carriers. Friction, as Tsing argues, is productive.[68] I have argued that opacity is likewise productive. By selectively highlighting meanings of jade as a symbol of national identity in China while shading the violent politics of jade mining in Myanmar's Kachin State, traders and scientific and popular dissemination in China contribute to what Proctor calls 'agnogenesis', the production of ignorance.[69] Opacity can serve a metaphor for what we cannot fully see and know; for example, a market on the other side of a national

67 Walsh, 'In the Wake', p. 234.

68 Tsing, *Friction*.

69 Proctor, 'Agnotology'.

border or a trade network we do not have access to. But in the trade of jadeite jade, opacity is also a concrete effect of the surface crust that hides the interior content of jade stones. As such, opaque jade surfaces serve as what Carey and Pedersen call 'infrastructures of doubt', but also, as I have noted, as material infrastructures for jade traders' economic strategizing in the form of gambling and attempts to reduce tax expenses.[70]

Attention to spaces between India and China explored in this volume shifts with geopolitical developments. Both Harris' discussion of how some trade routes and areas come into view, while others are obscured, and Pratt's argument about how the powerful definition of legitimate linguistic (or literate) interaction makes that which does not conform to this order invisible or anomalous, i.e., illegitimate linguistic interaction, are useful in considering shifting 'centralization' and 'peripheralization' of Burma-China borderlands.[71] Imports of jadeite as Imperial tribute in China's Qing Dynasty brought Kachin areas of northern Burma into connectivity with global capitalist systems and thereby into view of state agents. In turn, official restrictions on Burma-China trade after World War Two diverted jadeite flows and facilitated underground cross-border trade, but also to some extent peripheralized Burma-China borderlands as associated with lawlessness, in contrast to ideas of centre, civilization, progress, and law and order.[72] A 1994-2011 ceasefire between the KIA and the Tatmadaw was accompanied by increasing Chinese involvement in extraction and trade of natural resources like jade in Myanmar's Kachin State, facilitated by opaque deals with government and military agents. Belt and Road Initiative projects like the China-Myanmar Economic Corridor lay the infrastructure for further upscaling of resource extraction, trade, and contact along China's borders, contributing to what Rippa calls a 'centralizing of peripheries'.[73]

Realising the China-Myanmar Economic Corridor project through opaque business deals between political, military, and economic elites risks escalating frictions and conflicts between groups competing for resources, as upscaling of jade mining in Myanmar's Kachin State has done. It may further provide fertile ground both for conspiracy theories and xenophobia among dispossessed groups in Myanmar and for a 'blissful ignorance' about local consequences of China's global economic expansion among Chinese citizens

70 Carey and Pedersen, 'Introduction'.

71 Harris, *Geographical Diversions*; Pratt, 'Arts', p. 38.

72 See Chang, *Beyond Borders*, pp. 12-13.

73 Rippa, 'Centralizing Peripheries'.

that is further fed by nationalism and censorship. While opacity lends itself as a productive social and material phenomenon and field of inquiry, greater transparency in Myanmar's jade industry could help mitigate some of its detrimental effects in Kachin State and facilitate more fair distribution of its revenues. The same argument seems relevant for other sectors of resource extraction and trade that are currently effectuated under the auspices of China's Belt and Road Initiative.

References

Carey, Matthew, and Morten Axel Pedersen. 'Introduction: Infrastructures of Certainty and Doubt'. *The Cambridge Journal of Anthropology* 35, no. 2 (2017): pp. 18–29.

Chang, Wen-Chin. 'From a *Shiji* Episode to the Forbidden Jade Trade during the Socialist Regime in Burma'. In *Chinese Circulations: Capital, Commodities, and Networks in Southeast Asia*, edited by E. Tagliacozzo and W-C. Chang, pp. 455–480. Durham, NC and London: Duke University Press, 2011.

Chang, Wen-Chin. *Beyond Borders: Stories of Yunnanese Chinese Migrants of Burma*. Ithaca, NY and London: Cornell University Press, 2014.

'Digging deeper: Inside the Hpakant Jade Mine'. *Myanmar Times*, 8 June 2016.

Egreteau, Renaud. 'The Burmese Jade Trail: Transnational Networks, China, and the (Relative) Impact of International Sanctions on Myanmar's Gem Industry'. In *Myanmar's Transition: Openings, Obstacles and Opportunities*, edited by N. Cheesman, M. Skidmore, and T. Wilson, pp. 89–116. Singapore: Institute of Southeast Asian Studies, 2012.

Frey, Robert, and Bertil Lintner. 'Jade in Burma. The Major Jadeite Source'. In *Jade*, edited by Roger Keverne, pp. 267–272. New York, NY: Van Nostrand Reinhold, 1991.

Global Witness. *Jade: Myanmar's "Big State Secret"*. London: Global Witness Limited, 2015.

Graeber, David. *Toward an Anthropological Theory of Value*. New York: Palgrave, 2001.

Harris, Tina. *Geographical Diversions: Tibetan Trade, Global Transactions*. Athens, GA: University of Georgia Press, 2013.

Hughes, Richard W. 'Burma's Jade Mines. An Annotated Occidental History'. *Lotus Gemology* (1999), accessed 2 December 2021, http://www.lotusgemology.com/index.php/library/articles/283-burma-s-jade-mines-an-annotated-occidental-history.

Humphrey, Caroline. 'Introduction: Trusting and Mistrusting Across Borders'. In *Trust and Mistrust in the Economies of the China-Russia Borderlands*, edited by Caroline Humphrey, pp. 9–37. Amsterdam: Amsterdam University Press, 2018.

Jones, Lee. 'Understanding Myanmar's Ceasefires: Geopolitics, Political Economy and State-Building'. In *War and Peace in the Borderlands of Myanmar: The Kachin Ceasefire, 1994-2011*, edited by Mandy Sadan, pp. 95-113. Copenhagen: Nias Press, 2016.

Kachin Development Network Group (KDNG). *Blood Amber: Military Resource Grab Clears out Indigenous Peoples in Kachin State's Hugawng Valley* (2019), accessed 2 December 2021, https://kdng.org/wp-content/uploads/2019/08/BloodAmber_English.pdf.

Kiik, Laur. 'Conspiracy, God's Plan and National Emergency: Kachin Popular Analyses of the Ceasefire Era and its Resource Grabs'. In *War and Peace in the Borderlands of Myanmar: The Kachin Ceasefire, 1994-2011*, edited by Mandy Sadan, pp. 205–235. Copenhagen: Nias Press, 2016.

Leach, Edmund. *Political Systems of Highland Burma: A Study of Kachin Social Structure*. London: The Athlone Press, 1970 [1954].

Levy, Adrian, and Cathy Scott-Clark. *The Stone of Heaven: The Secret History of Imperial Green Jade*. London: Weidenfeld and Nicholson, 2001.

Miller, Arthur, and Michael Davis. *Intellectual Property: Patents, Trademarks and Copyright.* Minnesota, MN: West Publishing Company, 1983.

Møller, Henrik Kloppenborg. 'Becoming Real Businessmen? Negotiating Huangniu Identities at a Fake Commodity Market in Shanghai', MSc. diss., University of Copenhagen, 2010.

Møller, Henrik Kloppenborg. 'The Potentials of *Feicui*: Indeterminacy and Determination in Human-Jade Interactions in South-West China'. In *Emptiness and Fullness: Ethnographies of Lack and Desire in Contemporary China*, edited by S. Bregnbæk and M. Bunkenborg, pp. 120–139. New York, NY and Oxford: Berghahn Books, 2017.

Møller, Henrik Kloppenborg. 'Spectral Jade: Materiality, Conceptualisation, and Value in the Myanmar-China Jadeite Trade', PhD diss., Lund University, 2019.

Naylor, Thomas. 'The Underworld of Gemstones: Part I – Under the Rainbow'. *Crime, Law & Social Change* 53, no. 2 (2010): pp. 131–158.

Notar, Beth E. 'Authenticity Anxiety and Counterfeit Confidence: Outsourcing Souvenirs, Changing Money, and Narrating Value in Reform-Era China'. *Modern China* 32, no. 1 (2006): pp. 64–98.

Openjadedata.org. 'Even with New Data, Valuing Myanmar's Jade Industry Remains a Challenge', accessed 2 December 2021, https://openjadedata.org/stories/how_much_jade_worth.html.

Osburg, John. 'Consuming Belief. Luxury, Authenticity, and Chinese Patronage of Tibetan Buddhism in Contemporary China'. *HAU: Journal of Ethnographic Theory* 10, no. 1 (2020): pp. 69–84.

Pedersen, Morten Axel, and Bunkenborg, Mikkel. 'Roads that Separate: Sino-Mongolian Relations in the Inner Asian Desert". *Mobilities*, 7, no. 4 (2012): pp. 555-567.

Pratt, Mary Louise. 'Arts of the Contact Zone'. *Profession* (1991): pp. 33–40.

Proctor, Robert. 'Anti-Agate: The Great Diamond Hoax and the Semiprecious Stone Scam'. *Configurations* 9, no. 3 (2001): pp. 381–412.

Proctor, Robert. 'Agnotology: A Missing Term to Describe the Cultural Production of Ignorance (and Its Study)' In *Agnotology: The Making and Unmaking of Ignorance*, edited by R. Proctor and L. Shiebinger, pp. 1–33. Stanford, CA: Stanford University Press, 2008.

Rippa, Alessandro. 'Centralizing Peripheries: The Belt and Road Initiative and its Role in the Development of the Chinese Borderlands'. *International Journal of Business Anthropology* 7, no. 1 (2019): pp. 1–22.

Rippa, Alessandro. *Borderland Infrastructures: Trade, Development and Control in Western China*. Amsterdam: Amsterdam University Press, 2020.

Sadan, Mandy. *Being and Becoming Kachin: Histories Beyond the State in the Borderworlds of Burma*. A British Academy Postdoctoral Fellowship Monograph. Oxford: Oxford University Press, 2013.

Sadan, Mandy. 'Introduction'. In *War and Peace in the Borderlands of Myanmar: The Kachin Ceasefire, 1994-2011*, edited by Mandy Sadan, pp. 1–28. Copenhagen: Nias Press, 2016.

Safonova, Tatiana, István Sántha, and Pavel Sulyandziga. 'Searching for Trust: Indigenous People in the Jade Business'. In *Trust and Mistrust in the Economies of the China-Russia Borderlands*, edited by Caroline Humphrey, pp. 205–229. Amsterdam: Amsterdam University Press, 2018.

Schumann, Walter. *Gemstones of the World*. New York: Sterling, 1976.

Sigley, Gary. 'Tea and China's Rise: Tea, Nationalism and Culture in the 21st Century'. *International Communication of Chinese Culture* 2, no. 3 (2015): pp. 319–341.

Smith, Martin. 'Reflections on the Kachin Ceasefire: A Cycle of Hope and Disappointment'. In *War and Peace in the Borderlands of Myanmar: The Kachin Ceasefire, 1994-2011*, edited by Mandy Sadan, pp. 57–91. Copenhagen: Nias Press, 2016.

Sun, Laichen. 'From Baoshi to Feicui: Qing-Burmese Gem Trade, c. 1644-1800'. In *Chinese Circulations: Capital, Commodities and Networks in Southeast Asia*, edited by E. Tagliacozzo and W-C. Chang, pp. 203–221. Durham, NC and London: Duke University Press, 2011.

Tsing, Anna Lowenhaupt. *Friction: An Ethnography of Global Connection*. Princeton, NJ: Princeton University Press, 2005.

Vann, Elizabeth. 'Production Matters: Consumerism and Global Capitalism in Vietnam'. In *Research in Economic Anthropology: Anthropological Perspectives*

on Economic Development and Integration, vol. 22, edited by N. Dannhaeuser and C. Werner, pp. 225–257. London: Emerald Group Publishing, 2003.
Walker, Jill 'Jade: A Special Gemstone'. In *Jade*, edited by R. Keverne. New York: Springer.
Walsh, Andrew. 'In the Wake of Things: Speculating in and about Sapphires in Northern Madagascar'. *American Anthropologist* 106, no. 2 (2004): pp. 225–237.
Woods, Kevin. 'Ceasefire Capitalism: Military-Private Partnerships, Resource Concessions and Military State Building in the Burma-China Borderlands'. *Journal of Peasant Studies* 38, no. 4 (2011): pp. 747–770.
Woods, Kevin. 'Conflict Timber along the China-Burma Border: Connecting the Global Timber Consumer with Violent Extraction Sites'. In *Chinese Circulations: Capital, Commodities, and Networks in Southeast Asia*, edited by E. Tagliacozzo and W-C. Chang, pp. 480–507. Durham, NC and London: Duke University Press, 2011.
Woods, Kevin. 'The Commercialisation of Counterinsurgency: Battlefield Enemies, Business Bedfellows in Kachin State, Burma'. In *War and Peace in the Borderlands of Myanmar: The Kachin Ceasefire, 1994-2011*, edited by Mandy Sadan, pp. 114–145. Copenhagen: Nias Press, 2016.
Yang, Bin. *Between Wind and Clouds: The Making of Yunnan (Second Century BCE to Twentieth Century CE)*. New York, NY: Columbia University Press, 2009.
Yu, Ming. *Chinese Jade*. Cambridge: Cambridge University Press, 2009.
Zhang, Jinghong. *Puer Tea: Ancient Caravans and Urban Chic*. Seattle, WA and London: University of Washington Press, 2014.

About the author

Henrik Kloppenborg Møller, Carlsberg Visiting Research Fellow at Global Sustainable Development, University of Warwick, UK. Møller's research interests focus on Chinese society and China-Myanmar borderlands, ethnic minorities, markets, cosmology, development, and anthropological theory. His publications include 'Borderlines, Livelihood, and Ethnicity in the Yunnan-Myanmar Borderlands: A Rohingya Jade Trader's Narratives' (2021).

9 Multiple Identities of Young Sittwe Muslims and Becoming Rohingya

Tharaphi Than and Htoo May

Abstract

As places of transition, corridors shelter species – including humans – that flow from one place to another. Humans in transition can host multiple identities rather than affirming a singular identity associated with a singular space, race, or religion. This chapter argues that 'Rohingya-ness' is born along with mobility in defiance of fixed spaces and identities. The identities of Sittwe Muslims metamorphose, mutate, and camouflage as they physically and socially transition in and out of their chosen and forced places and confinements. Their multiple identities other than Rohingya challenge colonial ethnographic studies as well as the imaginings of champions of global rights as to who the Rohingyas are. The Rohingya identity of Sittwe Muslims has been strengthened by their mobile state, physically and socially.

Keywords: Sittwe Muslims, Rohingya, identity, belonging, Arakanese, Myanmar

Terms such as 'zero-line' are common when referring to borders.[1] Countries are visualized as having fixed beginning and end points. In turn, fixed physical boundaries create mental borders between people of different races and religions. Geography precedes identity, and fixed boundaries make fixed identities possible. David Ludden argues that 'territory is a project of social power striving to organize, direct, or contain mobility'.[2] Mobility of young Muslims in Sittwe of Rakhine State, Myanmar, preceded attempts by different actors – be they state(s), national or international agencies as

1 'It is finally clear'.

2 Ludden, chapter 2 in this volume, p. 31

Cederlöf, Gunnel, and Willem van Schendel (eds), *Flows and Frictions in Trans-Himalayan Spaces: Histories of Networking and Border Crossing*. Taylor & Francis Group 2022
DOI: 10.5117/9789463724371_CH09

well as ordinary citizens – to prescribe or deny them a certain identity, i.e., Rohingya. Their identity facilitated by their forced physical mobility and voluntary social mobility faces 'a brutal intervention to regulate it'.[3]

The term 'corridor' in this volume 'encompasses fluctuating and shifting – even volatile – movements across an often simultaneously changing socio-political geography'.[4] This concept of corridor needs not be limited to a physical space. It can be conceptually expanded to include identity and belonging to challenge the fixed notion of borders, identities, and belonging. The corridor metaphor blurred the rigid conceptualization of spaces. As a transition place, corridors shelter species, humans included, that flow from one place to another, even in formidable terrains such as the Eastern Himalayas – a focal point of this volume.[5] Like corridors, humans in transition or at an intersection of their lives for various reasons – be they political, economic, religious, or environmental – could host multiple identities rather than affirming to a singular identity associated with a singular space, race, or religion.

This chapter will argue that 'Rohingya-ness' is born along with mobility in defiance of attempts to closely link spaces and identities. Born and raised in Sittwe, Rohingyas or Sittwe Muslims are in fact Sittwe Arakanese first; trapped in the camps,[6] they become Rohingyas; when they engage with global pop cultures, they become K-pop fans or stans. Sittwe Muslims embody the concept of mobility, corridors, and flows. Their identities metamorphose, mutate, and camouflage as they physically and socially transition in and out of their chosen and forced places and confinements. More importantly, their multiple identities other than Rohingya challenge the colonial ethnographic studies that tie certain 'tribes' to particular regions[7] as well as challenging global rights champions' imagining of who Rohingyas are.[8] For Sittwe Muslims, identity is not tied to a fixed physical space. Contrary to the popular notion that associates Rohingyas with a specific physical space between coastal regions of Rakhine to mountain

3 Van Schendel and Cederlöf, chapter 1 in this volume, p. 17.

4 Ibid., p. 13.

5 Ibid.

6 Some Muslims live in 'host' villages, buying or sharing land or a house with existing Rohingya families there. But since they are not allowed to leave their villages, I will use the term 'camp' to reflect their camp-like conditions.

7 Van Schendel and Cederlöf, chapter 1 in this volume, p.14.

8 For how Rohingyas identify with 'Rohingya-ness' or refuse Rohingya identity, see Prasse-Freeman, 'Refusing Rohingya'.

ranges in Chittagong, the Rohingya identity of Sittwe Muslims has been strengthened by their mobile state, physically and socially.

Orientation to names and places

The focus of this paper, Sittwe Muslims, particularly a small group of young Sittwe Muslims who have been taking English lessons with me as their teacher since October 2018, came from Sittwe, the current capital of Rakhine State in Myanmar. As shown on the map, Sittwe is located close to the north of Rakhine. The majority of Rohingyas live in the two northernmost towns in Rakhine, Buthidaung and Maungdaw, but Sittwe, a town of nearly 150,000 people, also is home to many Rohingyas.[9] According to the Rakhine Commission formed after the 2012 interracial riots in Rakhine, Sittwe suffered the worst during the June riots, with more than 3,600 Rohingyas' homes being destroyed.[10] The students lived in Rupa South Ward, a few hundred metres from the downtown clocktower, and were forced to relocate to the outskirts of Sittwe, a place they called *kyuns* or islands, i.e., sparsely populated villages. After the riots, downtown Sittwe no longer has any Muslim inhabitants – neither Rohingyas nor Kamans[11]. Sittwe is now completely segregated unlike in the pre-riots era; Buddhist Arakanese live in the centre of the town surrounded by villages and newly established camps inhabited by local and displaced Rohingyas from other towns in Rakhine. Sittwe Muslims from the downtown area were considered economically well-off compared to other Rohingyas from the north or coastal areas such as Kyauk-phyu. The former were business people and governmental employees, whereas the latter engaged in fishing and farming, often as wage labourers.

My students' houses in central Sittwe were torn down during the sectarian riots in June 2012, and one student family decided to live in a host village rather than in camps where most Rohingyas were forced to relocate to.[12] All

9 Due to controversies over ethnicity or nationality, the 2014 census did not publish the breakdown of races or ethnicities.

10 Rakhine Conflicts Investigation Commission, p. 30

11 Kamans are counted as the state 'official' minority, whereas Rohingyas are not.

12 Some Rohingyas live in the shelters built by the government. Some, like Htoo May's family, bought the land inside the Rohingya compound. Though the shelters were constructed as temporary structures, no families could return to their old houses after seven years. In fact, many were forced to sell their remaining properties back in Sittwe for fear of them being confiscated. It is almost certain that these Rohingyas will never be able to go back to their places of origin, with the government claiming it is safer for them to remain in camps.

the Rohingyas from Sittwe and beyond have to remain inside camps, and those who have attempted to leave the camp to pursue their education or for other reasons without proper identification cards and special permission from the Ministry of Labour, Immigration and Population (La-Wa-Ka in Burmese) have been sent to prison or had to pay extortionate bribes if caught.[13] Most people in the camps and villages rely on food handouts from international donor agencies. Temporary schools for children were established, and many Rohingyas, especially those attending universities before the riots, became teachers in these schools.

Methods

Scholars working on borderlands or borderworlds[14] are acutely aware of the paucity of 'common' voices in their field of work. Two of the methodological challenges that decolonizing scholars grapple with are 'how does one centre the periphery?' and 'how does one (re)present the most invisible in the periphery?' Sittwe Muslims in the camps are included in this periphery category;[15] they are often referred to as part of the larger Rohingya group. Yet, their close proximity to the present-day capital of Rakhine State, Sittwe, gives them a distinct Arakanese identity, many of them speaking Arakanese as their first language. The challenge this chapter undertakes is to unpack the assumption that they are, and identify as, Rohingyas. However, rather than resorting to the ongoing debate to arbitrate who Rohingyas really are, this chapter is an attempt to bifurcate 'Rohingya' identity, to demonstrate the 'lumpiness' of identities,[16] and how young people switch in and out of a Rohingya identity on a day-to-day basis in their multifaceted lives online and offline. It is also an initiative to write about Rohingyas without essentializing, exoticizing, and unintentionally othering them by privileging the imagined, universal Rohingya identity.[17]

The research in this chapter used ethnography by proxy as a method. Proxy here denotes two substitutes or bypassing: 1) bypassing the exact

13 Myat Mon, 'Muslim Woman'.

14 See Sadan, *Being and Becoming*.

15 Writing about Rohingyas is a risky undertaking for Burmese scholars. Although many Burmese scholars and writers, especially those living in Myanmar, have experienced and know first-hand how the events unfolded, few research and write on the topic of Rohingya for fear of being ostracized. Ordinary Rohingyas like my students living in Myanmar amongst Rakhines and Burmans also fear for their lives if they speak out. Consequently, those living in freer countries and places speak for and write about Rohingyas. Subaltern Rohingyas might never be able to speak out.

16 Willem van Schendel suggested this terminology. Personal Communication, 21 November 2019.

17 Frydenlund, 'Motherhood, Home, and the Political Economy', 236-241.

location of the field, and 2) bypassing the exact topic of the research. I began teaching English to eight young Rohingyas aged 18 to 25 in 2018 through online portals such as Adobe Connect, Google, and Facebook. For nearly two years, I had a chance to observe the dynamics of this group of young people, and this chapter was conceived based on these observations. I will call this research method digital participant observation. No structured questions, interviews, or surveys were conducted. Even when I had a chance, I did not ask them questions about the violence they faced, so that they did not have to re-live their trauma for my research. All the observations in this chapter emerged from the discussions around English lessons with various topics ranging from Islam in Southeast Asia to K-pop.

There are advantages and disadvantages of using this method. The obvious advantage is the access to our 'subjects' through online teaching as we could not physically go and live in the immediate surroundings of the students. One author lives in the US, and even while she was in Yangon for one to seven months at a time, she was not officially allowed inside camps – although sneaking in and out would certainly be possible for a light-skinned Burman female like her. Female students have not been out of the camps and their host villages since 2012. Restricted mobility and access to the community are the major challenges for anthropological and sociological studies of Rohingya communities.[18] This explains the near-total absence of ethnographic studies of Rohingya and the disproportional reliance on the elite narratives of Rohingya lives and identity, mostly those living in exile.[19] On the other hand, although my students have restricted physical mobility, their social mobility is more fluid thanks to internet access.[20] And our common languages, Burmese and English, made our interactions possible. Language here is an enabler for research mobility and social mobility for the students.

Mobility and identities

After being forced out of downtown Sittwe, my students are currently known as IDPs (internally displaced people). Their lives as IDPs are more visible, more documented, and their information as a collective group of

18 Anthropological studies on Myanmar tend to focus on Buddhism including Weizzas and Nats. For more on Anthropology of Myanmar, see Schober, 'Belonging', p. 158.

19 Frydenlund, 'Motherhood, Home, and the Political Economy'.

20 During the ongoing civil war between the Arakanese Army (AA) and the Myanmar army, internet connections in many towns in Rakhine have been cut off. My students can, however, continue to access the internet.

IDPs is widely accessible through relief websites and humanitarian agencies. Ironically, their visibility is thwarted by their physical immobility, which is tied to them not being able to secure, and in most cases not wanting to secure, government-issued ID cards. These cards will categorize them as Bengalis rather than Rohingyas. For Rohingyas, if they object to 'Bengali' as their race/ethnicity,[21] they cannot apply for a National Verification Card (NVC). Rohingyas are not one of the '135' races[22] that the government recognized as 'official' races, and Rohingyas will be identified as 'Bengali' on the identification cards (ID 1, 2, and 3). Who is entitled to an ID card and consequently who is entitled to rights such as the right to education and the right to free movement within the borders of a country and beyond are all tied to one's racial/ethnic identity. As Ludden argues, mobility has been subjected to violent intervention by the state through the control over ethnicity.[23] There is a hierarchy of races in Myanmar, with the top ones being Burman followed by Mon, Rakhine, Kachin, Kayah, Karen, Chin, and Shan, ethnicities that were awarded their own designated states. But religion is deemed more important to determine one's entitlement to citizenship. If one is a Muslim, one cannot choose 'Bamar' or 'Rakhine', especially after the second wave of Rohingya exodus took place in 1992,[24] but one is forced to identify as 'Pakistani' Muslim, 'Kaman' Muslim, or 'Bengali' Muslim.

Caught up in the controversies over Rohingya identity, the students did not apply for ID cards known as National Verification Cards (NVC) until 2019. One student said 'we are not Bengalis and we do not feel close to anyone in Bangladesh. Why should we be identified as Bengalis?'[25] Not only the forced choice of countries but also where Rohingyas are from within Rakhine State is very important to the students. They said 'we speak Arakanese [Rakhine]

21 In the Burmese language, the term လူမျိုး covers race, ethnicity, and nationality. Its literal meaning is 'type of a person.'

22 An arbitrary number compiled by the British censuses based on languages people spoke in different parts of Burma. Though contentious as it leaves out many ethnic groups and repeats some groups, the Myanmar Government decided to use this number as their baseline for nationality. Any group not identified as one of the 135 must go through various verification processes. Rohingyas have to go through the most stringent of these as they must prove that their ancestors moved to Myanmar before 1826.

23 Ludden, chapter 2 in this volume.

24 Although one could be both a Bamar and Muslim in the past, after the 1982 citizenship laws came into force it is increasingly difficult to be both.

25 In July 2020, the family of one of the students received their IDs. They spent about USD 200 for each member of the family upon application and another USD 300 each when they obtained their IDs. Even after they submitted to being labelled as 'Bengali', Rohingyas still had to pay an extortionate amount of money to pass through all the bureaucratic procedures quickly, which effectively forces poor Rohingyas to remain ID-less and camp-bound.

Figure 9.1 Different ID cards in Myanmar[28]

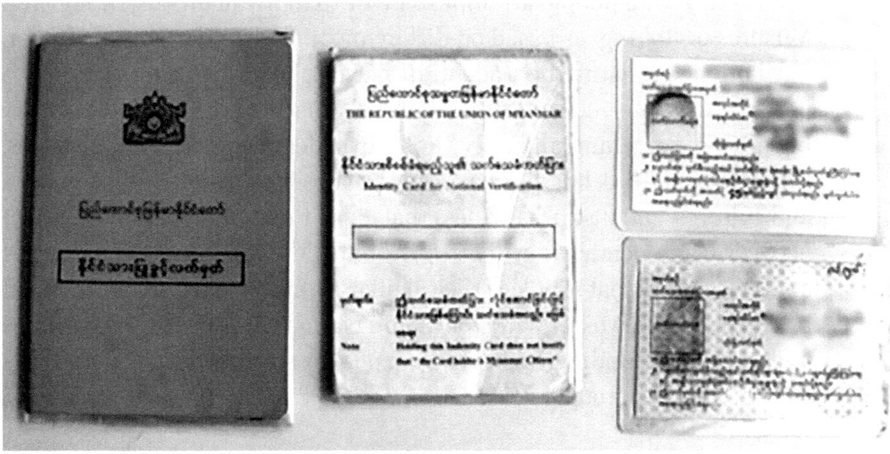

unlike them. We are people from downtown Sittwe, and they are people from the outskirts of Sittwe, which is more like the countryside. They do not speak Arakanese or Burmese. They can only speak our language.'

Lumping everyone together as Bengali or Rohingya robs young Muslims of their right to self-determination and freedom to assume and switch between many identities. The Myanmar Government assumes they are Bengalis. Rights groups assume they are Rohingyas. The reality is more nuanced. Rohingya identity, as we will argue later, is brought about by mobility between different groups of Rohingyas; for the students, Sittwe identity is associated with rootedness, security, and prosperity. Only when they were forced to move, did they attempt to identify as well as resist Rohingya-ness, albeit tacitly as evidenced in the words of the students.

Young Muslims of Sittwe are often seen through 'cultural' and 'ethnic' lenses that lumped them together with South Asian groups, particularly those in Banga.[27] Their bodies and movements have been highly regulated

26 Cards 1 and 2 say 'Identity Card for National Verification'. When the 'verification' process is complete, one is issued 3. Generally, Muslims, Hindus, and Christians are more likely to obtain ID 3 (after temporarily being granted ID 1). Buddhists will receive ID 4 without having to go through the verification process. Photo courtesy of Py, Aung. 'Will the race disappear after other races engulfed it or will it disappear because of the Ministry of Manpower and Immigration?'. The Chronicle, no. 69 (October 2019), pp. 8-16, p. 11.

27 Kyaw Minn Htin, an Arakanese scholar, uses the term 'Banga' to describe Rohingya communities that spread over Northeast India, Bangladesh, and Myanmar. Htin, *Where Mandalas Overlap*.

and collectively they are seen as Muslim others and a threat to Myanmar.[28] Yet, these young people attempt to pivot to many identities other than Muslims or Rohingyas based on different situations that they encounter. The Rohingya identity the students displayed along with other identities is part of a sensitive and carefully calibrated process in the midst of competing powers that claim and reject Rohingya identity. Their lived histories, learned experience, linguistic expertise, exposure to international politics and cultures, and digital savviness that allow them to adopt and negotiate between many identities have been obscured by the popular media narratives and scholarship that privilege the singular 'Rohingya' identity represented by elite Rohingya voices[29] and the 'Rohingya' name associated with images and terms surrounding genocide, boat people, stateless minority, and the most persecuted group.[30]

Adjacent identities and dangers of not belonging

As Duncan McDuie-Ra argues, minorities living in ethnic and geographical corridors of India-China often assume multiple identities. McDuie-Ra highlights how Mizo in Northeast India displays 'adjacent identities', i.e., pan-Northeast identity and pan-Mizo identity in the response to racism, to express solidarity, and to collectively partake of cosmopolitanism and global cultures through Mizo-dubbed Korean soap operas, Chinese action films, and Hollywood blockbusters. Proximity to specific issues such as racism and regions such as Delhi allows minorities in Northeast India to adjust their self-chosen identifications.[31] What circumstances enable or facilitate these adjacent identities warrant further exploration, especially among Rohingyas for whom choosing only one identity, i.e., Rohingya, poses both political risks and risks in day-to-day activities.

Willem van Schendel writes 'In South Asia the category of "tribe" is an undeniably important tool of identity politics.'[32] The Northeast India tribes

28 For Islamophobia, see Kyaw, 'Islamophobia', pp. 183-210. For othering Rohingyas, see Wade, *Myanmar's Enemy Within.*

29 Prasse-Freeman, 'Refusing Rohingya'.

30 A popular headline reads 'Who are the Rohingya? More than one million Rohingya Muslims are described as the "world's most persecuted minority"'. *Aljazeera*, 18 April 2018, accessed 30 July 2020, https://www.aljazeera.com/indepth/features/2017/08/rohingya-muslims-170831065142812.html.

31 McDuie-Ra, 'Adjacent identities', pp. 400-413.

32 van Schendel, 'Dangers', pp. 19-41, at 25.

Figure 9.2 U Ba Tun, a Rohingya broadcaster, presenting Rohingya programmes at the Burmese Broadcasting Station[35]

မြန်မာ့အသံ၌ တိုင်းရင်းသားရိုဟင်ဂျာဘာသာ အစီအစဉ်မှူး
ဦးဘထွန်း (B.A, B.L) အသံလွှင့်စဉ်ပုံ

must use their tribal status to make claims. Similarly, Rohingyas must make their rights claims in the language of their ethnicity and indigeneity for citizenship.[34] As South Asian, especially Northeast Indian, scholars continue to use the term 'tribe'; scholars of Rohingya continue to use the term

33 Original caption: 'Head of Rohingya Taiyintar Language Program (U Ba Tun, BA, BL)'. Maung Zarni, 'Official evidence'. This is an example of textual, visual, and auditorial legitimacy. The picture is of a Rohingya broadcasting programme manager, U Ba Tun, who had a Bachelor's degree in arts and law, in a recording studio at the Burmese Broadcasting Station in Rangoon in the 1960s. States endorsing Rohingya-ness and accepting Rohingyas have previously been used by Rohingyas to prove their legitimacy.
34 Ibid., p. 24.

'Rohingya' to refer to all refugees. The term and powerful discourses around this term that can decide the fate of Rohingyas put tremendous pressure on Arakanese-speaking Muslims with various labels such as 'Rohingya' and 'Bengalis'. Often caught up in what van Schendel calls 'claim research',[35] dangers of not-belonging (to Rohingyas) outweigh dangers of belonging for Rohingyas living in the camp. But as I will describe in the next section, risk accounting based on these identities, i.e., between Arakanese and Rohingya, fluctuates, and there are equally important identities such as Sittwe and K-pop identities in between two dominant identities.

Identities corridor

Out of eight students, one male student preferred to be known by his Rohingya name starting with 'Ro'. One of his friends, a female student, said during the conversation about names 'You are "Myo Aung",[36] not "Ro X"'. Her comment highlighted Arakanese Rohingyas becoming Rohingya by adopting Rohingya names. We observed that this practice is more common among Rohingya males than females. Female students would choose English names for their online profiles and they often used Korean K-pop female stars and pictures of their idols such as Malala Yousafzai as their profile icons. Names do not necessarily always signify their ethnic allegiance. By being able to switch from one name to another online shows their flexibility, temporal preferences, and being in control of their agency – at least in the virtual sphere.

Female students see themselves more as residents of Sittwe or Sittwethu[37] than Rohingyas. As Arakanese speakers, and particularly a dialect of Arakanese spoken in the Arakan capital, Sittwe, they distinguish themselves from other Rohingyas in the camp who know no Arakanese. They frequently use the pronouns 'us' and 'them' to distinguish Sittwe Rohingyas from others, particularly those from the Northern towns who cannot speak Arakanese. During class discussions, they hardly used the pronoun 'Rohingya', nor did they use the term in their online poems.

Submitted for the 2020 International Women's Day, one female student used the same analogies for being a woman and being a Rohingya. She wrote

35 Ibid., p. 26.

36 Pseudonym.

37 'Thu' and 'Thar' suffixes are used to describe female and male residents of a particular place. Sitthwethu means female residents of (and those who are born in) Sittwe.

a similar poem in Burmese towards the end of 2019 to describe her life as a Rohingya.[38] Although she submitted the poem under her pseudonym, she chose not to use the term 'Rohingya' in it. She usually referred to herself as 'We Muslims' in our conversations.[39] As Mandy Sadan writes about Kachins and other variant terms denoting them, they exercise a self-reflective practice to not use the term 'Rohingya'.[40] The poem reads:

> We're not birds,
> Don't put us in the cage.
> We're not robots,
> Don't control our feelings.
> When we are born,
> Don't be sad and frustrated.
> We don't have wings,
> But we have millions of dreams,
> Give us a chance.
> We have our own emotions,
> Be supportive to us,
> We have the strongest fortitude.
> We're girls,
> Don't kill us repeatedly with that word.
> We're ourselves and are women.[41]

For them, writing poems is a performative of an oppressed being with suppressed feelings and identities. Regarding statelessness and precarity, Judith Butler argues that 'performativity has everything to do with "who" can become produced as a recognizable subject, a subject who is living, whose life is worth sheltering and whose life, when lost, would be worthy of mourning'.[42]

Through poems, the students achieve visibility to others, i.e., others outside the physical and social confines of being IDPs. They can also explore

38 Sarar, 'Hopeless Cage'.

39 Non-Rohingya Muslims from Myanmar, however, attempted to distance themselves from Rohingyas for fear of collective persecution. On 27 November 2017, the All Myanmar Moulvi (Ulama Al Haqe) Organization of Myanmar issued a request to undocumented citizens, a euphemism for Rohingyas, to apply for National Verification Cards (NVCs), equivalent to a national ID. 'Five Muslim Groups Urged to accept NVC', *BBC Burmese*, accessed 31 July 2020, https://www.bbc.com/burmese/42256537.

40 Sadan, *Being and Becoming*, p. 6.

41 Sarar, 'Wish'.

42 Butler, 'Performativity', pp. i-xiii.

other intellectual identities such as a poet, the identity they often keep secret from their families.

As Nyamnjoh highlighted, 'Fiction is one of the most common vehicles used by African intellectuals to document and share "insider" accounts of their societies.'[43] Poems such as those by my students are increasingly popular on Facebook groups such as 'The Art Garden Rohingya' and provide uncensored accounts of their lives. Compared to exiled Rohingyas living in the US, Canada, and Europe, my students' suppressed pronouns and their Rohingya identity highlight the unequal power to use the term 'Rohingya' among different Rohingya communities.

Being a Muslim and being a Sittwe Arakanese who can speak and teach Arakanese to others is their most comfortable or public identity that the students used in class, in our conversations. One of the students wrote about double persecution she faced as a Rohingya and as a woman.[44] When asked about male students who were absent, sometimes for weeks, the students explained that male Rohingyas in the camp have more mobility than women. They can travel to other places via village routes, often for religious reasons, whereas female Rohingyas cannot leave the camp.

Before they lived in the camp, the students rarely used the Rohingya language, since they mostly spoke Arakanese at home and outside. They see Rohingyas from the outskirts of Sittwe as less academically able. The latter's lack of Arakanese language skills was seen as a barrier to education which is Burmanized and Bamar-centric. The students are often self-employed, giving private tuition at home to Rohingya students. They use the Rohingya language to communicate with their students, but perhaps with the hope of everyone getting out of the camp soon, they teach Burmese and Arakanese to those who do not know these languages. They said that their command of Rohingya is better inside the camp, although they had to learn words such as 'heart' and numbers greater than ten in Rohingya after they had moved to the camp. Their camp life brought them closer to other Rohingyas and they began to acknowledge that Rohingya was their language, like that of their ancestors. It is through this newly shared language that they have come to realize to which groups they can belong.

43 Barber, 'Introduction', pp. 1-12; Newell, 'Constructions', pp. 169-188; Newell, *West African Literatures*; Sullivan, 'Redefining the Novel', pp. 177-188.

44 For double exclusion as a Rohingya woman and as a productive workforce member, see Frydenlund and Lei, 'Double exclusions'.

Failures of civic institutions

The Arakanese language affords the students an Arakanese identity, which they are clearly proud of. Yet, they are denied this identity by the rest of Arakanese-speaking (non-Rohingya) people, since they are not yet allowed to re-enter the Arakanese-speaking world. Eight years after the sectarian riots in 2012, they are still trapped in camps. Only through online interactions did they display Arakanese identity as an erstwhile member of a vibrant community from downtown Sittwe.

Arakanese-speaking Rakhine people claim Sittwe as theirs, thereby robbing Arakanese-speaking non-Rakhine people of a sense of belonging. Nearly two hundred years ago, the British census takers and ethnographers failed to grapple with the complexity of the Arakan region, and such a colonial legacy of categorizing people based on language and religion and demarcating boundaries based on these categories is woefully sustained. Formulistic accounting of ethnicities and 'scientific' categorization of races based on language have continued to rank and divide communities.

As Michael Charney writes

> 'It was only in 1826 that southern Rakhine was rejoined with northern Rakhine under the Bengal Presidency after a period of forty years of occupation by the Burmans. The British decided to make a singular, abstract "Rakhine," one that glossed over its more mundane geographical and ethnic features and was oriented around Buddhism and Myanmar. The British had decided that nations have the same culture, language, and religion.'[45]

As residents of the Rakhine-majority living in downtown Sittwe, the students were able to take part in community activities ranging from Buddhist carnivals after the monsoon Lent period to water festivals. They competed in games such as tug-of-war organized for different wards. They also talked about their childhoods and recalled joining various festivals organized around major religious days – be they Buddhist, Hindu, or Muslim. Community theatres, fairs, and sports competitions are active civic spaces where residents of all religions gathered together and participated in the mostly fun activities. One writer wrote of a Muslim lawyer in a small town in the dry zone who directed the Buddhist youth theatre for pagoda festivals in the

45 Charney, 'Short and Long Term', p. 5. For more on the British census takers' conceptualization of race, see McAuliffe, *Caste*.

early 1930s.[46] Such interracial and interreligious engagement and exchanges did not change until the last riots broke out in 2012.

Sittwe Rohingya young people reminisced about themselves following the funeral proceedings of a Buddhist monk. They also recalled seeing various Chinese snacks sold and distributed during the Chinese New Year. Different residents organized different religious festivals during which religious identities did not pose a barrier to sharing in the activities and feeling part of belonging to a larger community transcending one's race and religion. But after the June 2012 sectarian riots during which Muslims were targeted, the Islamic Council, the largest Muslim organization in Myanmar, issued a statement in April 2013 asking all Muslims to stay indoors during the biggest festival of the year in Myanmar, the Water Festival.[47] Civic space shrank, not just for Rohingyas but also for all Muslims of Myanmar after 2012.

At her elementary and middle schools, one of the female students was selected to represent her school in many local and regional competitions. She is aware that her good standing at school shielded her from discrimination that poor Muslim students were often subject to. She also came from a fairly rich family from downtown Sittwe where Rohingyas were a minority. She could pass as an Arakanese and not one of the 'Kalamas'. 'Kala' is a derogatory term used by many to refer to South Asian descendants as well as foreigners. Such an abusive word was then reserved for her less fortunate Rohingya schoolmates. Here the intersection of race and class is on display – Rohingyas from poor backgrounds who cannot attain good academic standing at school experienced more discrimination.

After finishing elementary school, the female student was taken to see the principal of a middle school[48] by her Rakhine teacher. Seeing her, the principal issued application forms. Only after the forms had been submitted and after she had been attending classes for a week, did the principal find out from her father's name that she was a Muslim. The principal let her stay, but she could have been rejected if she had not already been attending classes. Her physical appearance, particularly her fair skin, and her Burmese language skills gave the principal the impression that she was a Rakhine. Such factors should not be the criteria for eligibility to attend the school. When her older brother finished the middle school, the principal of the Basic Education High School Number 4 rejected his application. The principal

46 Thein, *Autobiography*, p. 34.
47 'Muslims asked to perform good deeds'.
48 It was Kyey-bin-gyi Middle School.

bluntly said, 'We cannot take the responsibility for your safety.'[49] He had no other options but to enrol in a Muslim-majority school.

Civic institutions like this high school have failed the students and many other Muslims, since there is institutional pressure, although it is implicit, not to accept Muslims, and individual school principals have submitted to that pressure. No one challenged such discrimination becoming the norm for Muslims, particularly Rohingyas. Regulation of Rohingya lives and bodies by the border security forces known as Na-Sa-Ka, the army, and the local authorities in Northern Arakan is well-documented,[50] whereas discrimination and exclusion by schools are little known, despite the latter impacting the mobility and mental health of young Rohingyas more severely. Failures of civic institutions to include Rohingya can even be seen as more important issues to address, but the assumption is that issuing 'citizenship' will fix everything.[51]

(Re)producing refugee identity and becoming Rohingya

There is a tendency to conceptualize that all the persecuted Rohingyas will affirm their 'Rohingya-ness' and rally around many global movements on Rohingya rights and their indigeneity, especially among Burman communities and media in Myanmar.[52] On the other hand, as van Schendel

49 In around 2011, there was a fight between Muslim and Rakhine students in Sittwe. This incident became an excuse for Rakhine-dominant schools not to enrol Muslims. However, fights between schools are not confined to groups with different religions. In the old capital, Rangoon, school fights used to be frequent, especially between rival schools with no religious affiliations.

50 Examples of regulations include monitoring and rationing the number of marriages among Rohingyas, monitoring and deciding on divorces, not registering births or issuing household registrations, and not allowing Muslims to use Burmese names. On 18 October 2018, the president abolished such practices by the Border Scrutiny Board (Na-Sa-Ka) restricting the rights of Rohingyas, but ironically many Rohingyas from Northern Arakan are now living outside Myanmar jurisdiction. These announcements highlighted that the government brought the issue of Rohingya under their direct control, abolishing different authorities including the defence/army and border control. The latter show that Rohingya lives are under the scrutiny of many different layers of governments, and such tight scrutiny has ripple effects on other government institutions such as schools. Although no laws or directives exist to ban Rohingyas from enrolling in schools, Rakhine-majority schools will not accept Rohingyas or Muslims, citing security reasons. *Government Gazette*, 18 October 2018.

51 Framing the citizenship issue as the root cause of Rohingya problems is common. See an example here: 'Human rights group says "citizenship scrutiny" has gradually limited Rohingya rights in Myanmar and at root of crisis.' Milko, 'Genocide card'.

52 Many Myanmar media highlight the narrative that the international lobby groups, media, and Rohingyas inside and outside the country rally together for the call for 'Rohingya' identity and to punish Myanmar. Aung Zaw, 'Diplomacy defense'.

notes, romantic activists and hard-nosed technocrats at various organizations, through their imagined Rohingya solidarity and aid programmes for refugees, make it harder for Rohingyas to not share this universal vision of Rohingya.[53] Two examples stand out.

Firstly, Rohingya students volunteer to be ward guards, funeral helpers, trustees, and teachers, and during the English lessons, we talked about how humanistic values such as generosity, kindness, and empathy allow people to help each other.[54] One student asked an intriguing question: 'How are these humanistic values different from human rights?' This student is working for an international NGO, and human rights language is the language that they are familiar with. Abandoned by most of the country and violently expelled from their houses by Rakhines, kindness and empathy are words through which they cannot relate to other communities. Engaged in pan-Rohingya-ism and daily humanistic work through his job, this student believed that human rights were the guiding principles for human interactions. For a Burman, Burmese-speaking educator, terms such as kindness, empathy, and sympathy, were taken for granted. However, for some communities such as the students' communities, these words have lost their meanings and practices based on the values attached to these meanings. The students are more familiar with institutional and legal languages such as rights than communal and vernacular words such as kindness and sympathy.

Secondly, a female student also works for a series of NGOs providing humanitarian assistance to Rohingyas. She has to apply for jobs, sometimes for the same organization, for the same position, after each project cycle ends. On her job applications, she has to emphasize that she is a refugee or an IDP, but she does not specifically say that she is a Rohingya. But as per the job application norms in Myanmar, she has listed her bio data such as religion and her father's name on her CV, and it is quite obvious from her CV that she is a Rohingya. Although non-Rohingyas can also apply for the same jobs, she has an advantage because of her command of the language. Her identity is her asset within the IDP community, and she is required to display it, albeit discreetly, before the beginning of every project cycle.

Through the language of the international organizations that emphasize human rights over humanistic values, Rohingya-ness is easier to imagine through the means of jobs and opportunities. Humanitarian organizations, which, with good intentions, privilege their language of rights, become

53 van Schendel, 'Dangers', pp. 19-41, at 24-26.

54 Loan words from Pali are used in the Burmese language: generosity is *cetana*, loving kindness is *metta*, and empathy is *karuna*.

more important players in the everyday lives of IDPs in the state-absent or state-neglected areas such as camps in Sittwe. NGOs are not only acting as pseudo-states but are also giving a sense of belonging to the imagined communities mobilized and mobilizing around human rights. When secular and civic organizations such as schools fail to foster civic identities among its students, the latter come to believe that the language and mechanisms of the Universal Declaration of Human Rights are their refuge and solution.

The same student who had to emphasize her powerlessness or placelessness by being an IDP (Rohingya) refused to apply for a national identity card (ID)[55] because she does not want to be called 'Bengali'. But she gave in after seven years of impasse between Rohingya communities and the government. In 2019, together with her family, she applied for an ID. If she eventually receives her ID, she will be known as a 'Bengali'. These everyday choices made by young Rohingyas highlight the fluidity and flexibility of their decisions. Identity is only proximate to immediate jobs and who they interact with online and offline. But governments, international organizations, and advocacy and rights group harden the plasticity of the identities that these young people want to assume.

Even citizenship cannot guarantee freedom of movement and other rights.[56] Citizenship or an ID card is not a solution as proposed by many. Nick Cheesman argued that Rohingyas (and I would include Chinese, Indians, and anyone not deemed 'indigenous' or 'taingyingtha') are not considered members of a political community and they are hence denied citizenship.[57] Belonging to one of the recognized 135 national races or ethnicities surpasses any other conditions for claiming citizenship,[58] and by being stripped of citizenship by the 1982 citizenship laws, Rohingyas do not have any recourse to restore their citizens' rights or the right to belong to and become active members of Burmese communities. The chair of the Democracy and Human Rights Party, the only registered political party mainly comprising Rohingyas, U Kyaw Min, argued that their identity and existence were suppressed by 'Arnar-saq', meaning 'state orders' or 'political power'.[59]

Through persecutions, especially after the 1982 citizenship laws came into force, Rohingyas are particularly vulnerable, since they are now considered visitors and not one of Burma's own peoples. Although Rohingya leaders

55 Also known as National Verification Cards (NVC) for Rohingyas or people who cannot prove their eligibility for citizenship.

56 Galache and Avezuela, 'Citizenship'. As quoted in Prasse-Freeman, 'Rohingya crisis', pp. 1-2.

57 Cheesman, 'How in Myanmar', pp. 461-483.

58 Ibid.

59 Interview with U Kyaw Min, Yangon, 8 November 2019.

such as U Kyaw Min believe that the state is the most powerful institution to redress their woes and restore their rights, social discrimination and religious intolerance often advocated by monks such as Sidaghu who have repeatedly said that there are two classes of people – hosts and visitors – sustain and even amplify unjust and dangerous laws. Social and religious mobilizations against Rohingyas also help the state to justify keeping the 1982 citizenship laws. While the state's laws are visible, social sanctions and prejudices are rather insidious and more difficult to tackle. And it is the latter that prompt U Kyaw Min and colleagues to amass power, since they have lost hope of appealing to their neighbours and non-Rohingya community members. Meanwhile, more powerless people such as the Muslim students of Sittwe have finally submitted to the state's demands, i.e., accepting NVC cards and being identified as 'aliens', 'to-be-naturalized citizens', or 'Bengalis'.

Cosmopolitan identities

Although they were once Sittwe Arakanese and have now applied for NVCs that will categorize them as 'Bengalis', the students freely express other identities online. One of them said that she and her friends became K-pop fans at school and organized small performances in class, imitating their favourite K-pop bands. She often used her K-pop idols as avatars on her Facebook profile and other communication apps such as Viber from Israel. All the students know Malala, the Nobel Peace Prize Laureate, and Rohingya icons such as Wai Wai Nu from Myanmar. They now write poems and submit material to larger Rohingya groups to commemorate days such as International Women's Day. They signed up for entrepreneur courses and list the schools on their profiles. They often change their Facebook names from Burmese to English and to Muslim ones. Their identities are certainly more fluid online.

They shared English lessons on Facebook as well as their favourite K-pop bands and their performances. While they displayed solidarity with pan-Rohingya movements by sharing news about Rohingya, they also had an extensive list of activities and many non-Rohingya identities online. Their friends in Yangon and further afield such as Ruili in China tagged them on their trips as a way to take them to different places – a wonderful virtual world outside their seemingly life-long imprisonment.[60] During the pandemic

60 For Rohingyas engaging in the jade trade in China and Myanmar, see Henrik Kloppenborg Møller, 'Frictions and Opacities in the Myanmar-China Jade Trade', chapter 8 in this volume.

in 2020, many online courses mushroomed in Myanmar, and the students enrolled in both free and fee-paying courses, sometimes attending two or three courses at the same time.

A turning point and future research

On 23 July 2020, a group of Muslim students donated funds to Rakhines displaced from their houses due to the ongoing war between the Arakan Army (AA) and the Myanmar army.[61] Many people shared such news as a ray of hope for two warring communities of Rakhines and Muslims (Rohingyas). In return, on 25 July 2020, Rakhine student union members visited Muslim (Rohingya) camps to show *metta*, loving kindness, and sympathy, words used by the visitors.[62] In the midst of intense fighting between the Myanmar army and the Arakan Army (AA), two groups of young students, Rakhine and Rohingya, found common ground as victims. Young people are now building bridges and expressing solidarity through common suffering. Words the Rohingya students did not hear when they were forced out of their homes, i.e., 'sympathy' and 'loving kindness' have been used by young Rakhines to show their feelings towards Rohingya. Rohingyas' collective bargaining for citizenship and the right to claim their identity is ongoing, yet initiatives by young people might prove to be a quicker route via which Sittwe Rohingyas could reclaim their robbed Arakanese identity and could openly display adjacent and discreet identities that they have hitherto reserved for the virtual world.

In terms of future research, ethnographic studies of how civic relations could be forged among Muslims and Buddhists (and others) will guide many stakeholders in how to avoid further harm caused by privileging one identity over another and how to let mobile young people imagine a variegated landscape of belonging. Such research includes looking at what could help them imagine themselves to be Sittwe Arakanese, Rohingyas, or cultural citizens of the world.

As argued here, the (forced) physical mobility and the dynamic social and cultural mobility of young Muslims enable them to imagine, assume, and inhabit multiple identities simultaneously. Languages that they share with their fellow camp residents, NGOs, and K-pop fans around the world

61 'Muslim students'.

62 Public Relations Committee, Arakan Students Union, *Facebook*, accessed 31 July 2020 https://www.facebook.com/asunion.tcom/?tn-str=k*F.

show that language is a powerful tool for mobility. Territorial controls, and racial, religious, linguistic, and global hegemonies through NGOs act as dams and stop these flows of identities and belonging.

One was not born with one's identity but acquired it through socialization and in the case of Rohingyas through state-imposed processes towards citizenship and persecution. Identity is a marker which should be acquired through an intimate, personal, and deliberate choice to affirm personhood, to celebrate one's family or families and history or histories, and to empower through (self)respect, but it becomes a weapon against oneself for Rohingyas. Both the state and civil society view Rohingya identity through a legal lens, and by declaring Rohingya identity illegal, Rohingyas are robbed of many other identities: neighbours, Sittwe residents, and even civil servants. As Mandy Sadan suggests,[63] ideological remodelling needs be considered. The concept of corridor rather than border will facilitate mutual recognition as opposed to seeking legitimacy in history; an emphasis on the flow and continuity of spaces rather than boundary markings will allow shared notions of civic and communal identities; and promoting the idea of belonging based on commonality rather than ethnicity and indigeneity will enable Sittwe Muslims to adopt many identities – be they Rohingyas, K-pop fans, or poets.

References

Aung Zaw. 'Is the Diplomacy Defense Broken?'. *The Irrawaddy*, 17 September 2017, accessed 30 July 2020, https://burma.irrawaddy.com/opinion/interview/2017/09/17/142777.html.

Barber, Karin. 'Introduction'. In *Readings in African Popular Culture*, edited by Karin Barber, pp. 1–12. Oxford: James Currey, 1997.

Butler, Judith. 'Performativity, Precarity and Sexual Politics'. *AIBR. Revista de Antropología Iberoamericana* 4, no. 3 (2009): pp. i–xiii.

Charney, Michael W. 'The Short and Long Term Roots of the Rohingya Crisis: The Eradication of a Myanmar Ethnic Group'. Cornell University Seminar and Roundtable on the Rohingya, 7 November 2017, Ithaca, NY.

Cheesman, Nick. 'How in Myanmar "National Races" Came to Surpass Citizenship and Exclude Rohingya'. *Journal of Contemporary Asia* 47, no. 3 (2017): pp. 461–483.

63 *Flows and Friction in Trans-Himalayan Spaces*, Conference organized by the India-China Corridor project, Stockholm, Sweden, 20-22 November 2019.

'Five Muslim Groups Urged to accept NVC'. *BBC Burmese*, accessed 31 July 2020, https://www.bbc.com/burmese/42256537.

Frydenlund, Shae A. 'Motherhood, Home, and the Political Economy of Rohingya Women's Labor'. In *Unraveling Myanmar's Transition*, edited by Pavin Chachavalpongpun, Elliott Prasse-Freeman, and Patrick Strefford, pp. 236–241. Singapore: NUS Press, 2020.

Frydenlund, Shae A., and S. Lei. 'Double Exclusions, Double Shifts: Muslim Women's Labor Subjectivities in Yangon, Myanmar'. *Independent Journal of Burmese Scholarship* (forthcoming).

Galache, Carlos Sardiña, and Antolin Avezuela. 'Citizenship for a few, rights for none'. *Equal Times*, 24 July 2017, as quoted in https://www.aljazeera.com/indepth/features/2017/08/rohingya-muslims-170831065142812.html, accessed 30 July 2020.

Htin, Kyaw Minn. 'Where Mandalas Overlap: Histories, identities and Fates of the People from Arakan and South-Eastern Bangladesh', PhD diss., National University of Singapore, 2017.

Interview with U Kyaw Min, Yangon, 8 November 2019.

'It is finally clear'. *News Eleven*, 26 February 2018, accessed 31 July 2020, https://news-eleven.com/features/40345.

Kyaw, Nyi Nyi. 'Islamophobia in Buddhist Myanmar: The 969 Movement and Anti-Muslim Violence'. In *Islam and The State in Myanmar*, edited by Melissa Crouch, pp. 183–210. Oxford: Oxford University Press, 2016.

Maung, Zarni. 'The official evidence of the Rohingya ethnic ID and citizenship which the Burmese ethno- and genocidists don't want you to see'. *Rohingya Blogger* http://www.rohingyablogger.com/2013/05/the-official-evidence-of-rohingya.html, accessed 31 July 2020.

McAuliffe, Erin L. 'Caste and the Quest for Racial Hierarchy in British Burma: An Analysis of Census Classifications from 1872–1931', MA diss., University of Washington, 2017.

McDuie-Ra, Duncan. 'Adjacent Identities in Northeast India'. *Asian Ethnicity* 17, no. 3 (2016): pp. 400–413.

Milko, Victoria. 'Genocide card: Myanmar Rohingya verification scheme condemned'. *Aljazeera*, 2 September 2019, accessed 30 July 2020, https://www.aljazeera.com/news/2019/09/genocide-card-myanmar-rohingya-verification-scheme-condemned-190903012922259.html.

'Muslim students from Sittwe donated money to Rakhine war victims'. *Narijara News*, 23 July 2020, https://burmese.narinjara.com/news/detail/5f194fe4700aa70862a0935c, accessed 31 July 2020.

'Muslims asked to perform good deeds', *DVB*, 9 April 2013, accessed 16 July 2020, http://burmese.dvb.no/archives/38299.

Myat Mon, Su. 'Muslim Woman from Rakhine Jailed for Attempted Yangon Travel', *Frontier*, 2 June 2018, accessed 26 October 2019, https://frontiermyanmar.net/en/muslim-woman-in-rakhine-jailed-for-attempted-yangon-travel.

Newell, Stephanie. 'Constructions of Nigerian Women in Popular Literatures by Men'. *African Languages and Cultures* 9, no. 2 (1996): pp. 169–188.

Newell, Stephanie. *West African Literatures: Ways of Reading.* Oxford: Oxford University Press, 2006.

Prasse-Freeman, Elliott. 'The Rohingya Crisis'. *Anthropology Today* 33, no. 6 (2017): pp. 1–2.

Prasse-Freeman, Elliott. 'Refusing Rohingya: Reformulating Ethnicity Amidst Blunt Biopolitics'. In *Current Anthropology* (forthcoming 2022).

Public Relations Committee, Arakan Students Union. *Facebook*, accessed 31 July 2020, https://www.facebook.com/asunion.tcom/?tn-str=k*F.

Rakhine Conflicts Investigation Commission. Press Release number 2, 5 November 2012, n.p.

Sadan, Mandy. *Being and Becoming Kachin: Histories Beyond the State in the Borderworlds of Burma*. A British Academy Postdoctoral Fellowship Monograph. Oxford: Oxford University Press, 2013.

Sarar. 'Hopeless Cage'. *Facebook: 'The Art Garden Rohingya*, accessed 29 July 2020, https://m.facebook.com/HeartbeatsOfRohingya/photos/a.1720068411463546/1720068738130180/?type=3.

Sarar. 'Wish'. *Facebook: 'The Art Garden Rohingya'*, accessed 28 July 2020, https://www.facebook.com/HeartbeatsOfRohingya/photos/a.1720068411463546/1790362767767443/?type=3

Schober, Juliana. 'Belonging in a New Myanmar: Identity, Law, and Gender in the Anthropology of Contemporary Buddhism'. *Religion and Society: Advances in Research* 8 (2017): pp 158–172.

Sullivan, Joanna. 'Redefining the Novel in Africa'. *Research in African Literatures* 37, no. 4 (2006).

Thein, Pe. *My Autobiography: Good and Bad*. Yangon: Electricity Publishing, 2009.

Van Schendel, Willem. 'The Dangers of Belonging: Tribes, Indigenous Peoples and Homelands in South Asia'. In *The Politics of Belonging in India: Becoming Adivasi*, edited by D. J. Rycroft and S. Dasgupta, pp. 19–43. London and New York, NY: Routledge, 2011.

Wade, Francis. *Myanmar's Enemy Within: Buddhist Violence and the Making of a Muslim 'Other'*, 2nd ed. London: Zed Books, 2017.

'Who are the Rohingya? The More than One Million Rohingya Muslims are Described as the "World's Most Persecuted Minority"'. *Aljazeera*, 18 April 2018.

About the authors

THARAPHI THAN, Associate Professor, Department of World Languages and Cultures, Northern Illinois University, USA. She studies the intersection of dissent and food, particularly rice, focusing on events causing the 1967 and 1974 Myanmar rice riots. Global collaboration with feminists will result in the volume *Field Feminism*. Publications include *Women in Modern Burma* (2014).

HTOO MAY, university student, poet and NGO worker from Sittwe, Rakhine State, Burma. She has lived in a camp for internally displaced people since 2012 when nationalists drove Rohingyas out of Sittwe. She teaches Rohingya children in her camp and contributes to Rohingya poetry blogs. She is a member of the Rohingya Student Union (Sittwe) and a volunteer Burmese-Rohingya translator.

Bibliography

Aberth, John. *Plagues in World History*. Lanham: Rowman and Littlefield, 2011.

Agarwalla, Haribilash. *Haribilash Agarwalla Dangariyar Atmajibani*. Guwahati: Tarunkumar Agarwalla, 1967.

Ahmad, Nisar. 'Assam- Bengal Trade in the Medieval Period: A Numismatic Perspective'. *Journal of the Economic and Social History of the Orient* 33, no. 2 (1990): p. 176.

Ahmed, Imdad Ali. 'Elephants in Medieval Assam'. *Proceedings of the Indian History Congress* 76 (2015): pp. 264–270.

Alikuzai, Hamid Wahed. *A Concise History of Afghanistan in 25 Volumes: Volume 14*. Victoria BC: Trafford Publishing, 2013.

Allan, J. 'The coinage of Assam'. *The Numismatic Chronicle and Journal of the Royal Numismatic Society* (1909): pp. 300–331.

Allen, B. C. *Assam District Gazetteer*, vol. 3, Lakhimpur. Shillong: Government Press, 1906.

Amrith, Sunil. *Crossing the Bay of Bengal: The Furies of Nature and the Fortunes of Migrants*. Cambridge, MA: Harvard University Press, 2015.

Anderson, John. *A Report on the Expedition to Western Yunan via Bhamo*. Calcutta: Office of the Superintendent of Government Printing, 1871.

Anderson, John. *Mandalay to Momien: A Narrative of the Two Expeditions to Western China of 1868 and 1875, under Colonel Edward B. Sladen and Colonel Horace Browne*. London: Macmillan, 1876.6

Appadurai, Arjun. 'Death Certainty: Ethnic Uncertainty in the Era of Globalization'. *Public Culture* 10, no. 2 (1998): pp. 225–247.

Appadurai, Arjun. *The Fear of Small Numbers: An Essay on the Geography of Anger*. Durham, NC: Duke University Press, 2006.

Assam, Government of. *Annual Report on the Administration of Land Revenue in Assam, 1874-75*. Shillong: Assam Secretariat Press, 1875.

Assam, Government of. *Report on the Administration of the Province of Assam for the years 1874-75 & 75-76*. Shillong: Assam Secretariat Press, 1877.

Assam, Government of. *Report on the Administration of the Province of Assam, 1876-77*. Shillong: Assam Secretariat Press, 1877.

Assam, Government of. *Report on the Administration of the Province of Assam, 1878-79*. Shillong: Assam Secretariat Press, 1880.

Assam, Government of. *Report on the Trade between Assam and the Adjoining Foreign Countries for the year 1881-82* [1882].

Assam, Government of. *Physical and Political Geography of the Province of Assam*. Shillong: Assam Secretariat Press, 1896.

Assam, Government of. *Report on the Trade between Assam and the Adjoining Foreign Tribes and Countries, 1924-25* [1925].

Atwill, David G. *The Chinese Sultanate: Islam, Ethnicity, and the Panthay Rebellion in Southwest China, 1856-1873*. Stanford, CA: Stanford University Press, 2005.

Aung Zaw. 'Is the Diplomacy Defense Broken?'. *The Irrawaddy*, 17 September 2017, accessed 30 July 2020, https://burma.irrawaddy.com/opinion/interview/2017/09/17/142777.html.

Aung-Thwin, Michael. *Pagan: The Origins of Modern Burma*. Honolulu: University of Hawaii Press, 1985.

Bae, Christopher J., Katerina Douka, and Michael D. Petraglia. 'On the Origin of Modern Humans: Asian Perspectives'. *Science* 358, eaai9067 (2017).

Barber, Karin. 'Introduction'. In *Readings in African Popular Culture*, edited by Karin Barber, pp. 1–12. Oxford: James Currey, 1997.

Bareh, Hamlet. *The History and Culture of the Khasi People*. Guwahati: Spectrum Publications, 1997.

Barpujari, Heramba Kanta. 'The Attempted Traffic with the Chinese through the North-East Frontier (1826-58)'. *Proceedings of the Indian History Congress* 24 (1961): pp. 326–331.

Barpujari, Heramba Kanta. *The Comprehensive History of Assam*, vol. IV. Guwahati: Publication Board, Assam, 1992.

Barua, Taz. 'Return of the Frontier: Understanding the Demands for Inner Line in Northeast India'. *Alternatives: Global, Local, Political* 42, no. 3 (2018): pp. 107–120.

Baruah, Sanjib. 'Citizens and Denizens: Ethnicity, Homelands, and the Crisis of Displacement in Northeast India'. *Journal of Refugee Studies* 16, no. 1 (2003): pp. 44–66.

Baruah, Sanjib. 'Politics of Territoriality: Indigeneity, Itinerancy and Rights in North-East India'. In *Territorial changes and territorial restructurings in the Himalayas*, edited by Joëlle Smajda, pp. 69–83. New Delhi: CNRS and Adroit Publishers, 2013.

Basic Statistics of North Eastern Region 2015, Government of India, North Eastern Council Secretariat, Shillong, accessed 2 December 2021, http://necouncil.gov.in/sites/default/files/uploadfiles/BasicStatistic2015-min.pdf

Baynes, Thomas Spencer. *Encyclopaedia Britannica: A Dictionary of Arts, Sciences, and General Literature, 9th edition*, vol. 12. Edinburgh: Adam and Charles Black, 1881.

Bengal, Government of. *Report on the Administration of Bengal, 1871-72*. Calcutta: Bengal Secretariat Press, 1872.

Bengal, Government of. *Report on the Administration of Bengal 1872-73*. Calcutta: Bengal Secretariat Press, 1873.

Bezbaruah, Ranju. 'Rubber Dispute of the Miripathar Daflas; British Attitude'. *Proceedings of North East India History Association* 14 (1993): pp. 157–162.

Bhattacharjee, Jayanta Bhusan. *Trade and Colony: The British Colonisation of North East India*. Shillong: North East India History Association, 2000.

Bhattacharya, Neeladri, and Joy L. K. Pachuau, eds. *Landscape, Culture and Belonging: Writing the History of Northeast India*. Cambridge: Cambridge University Press, 2019.

Bhattacharya, Rakhee. 'Neoliberal Developmentalism: State Strategy in India's North-East'. In *Developmentalism as Strategy: Interrogating Post-Colonial Narratives on India's Northeast*, edited by Rakhee Bhattacharya, pp. 169–191. New Delhi: Sage, 2019.

Bhuyan, Suryya Kumar. *Anglo-Assamese Relations, 1771-1826: A History of the Relations of Assam with the East India Company from 1771 to 1826*. Gauhati: Department of Historical and Antiquarian Studies, 1949.

Blackburn, Stuart. 'Memories of Migration: Notes on Legends and Beads in Arunachal Pradesh, India'. *European Bulletin of Himalayan Research* 25, no. 26 (2003): pp. 15–60.

Blackburn, Stuart. *Himalayan Tribal Tales: Oral Tradition and Culture in the Apatani Valley*. Leiden: Brill, 2008.

Blurton, T. Richard. *Krishna in the Garden of Assam*. London: British Museum, 2016.

Bora, Papori. 'Politics of Difference in the Northeast: A Feminist Reflection'. In *Northeast India: A Place of Relations*, edited by Y. Saikia and A. R. Baishya, pp. 244–261. Cambridge: Cambridge University Press, 2017.

Bose, Kishen Kant. 'Some Account of the Country of Bhutan', Translated by David Scott. *Asiatick Researches* 15 (1825): pp. 128–137.

Bose, Manilal L. *British Policy in the North-East Frontier Agency*. New Delhi: Concept, 1979.

Bowers, Alexander. *Bhamo Expedition: Report on the Practicability of Re-Opening the Trade Route, between Burma and Western China*. Rangoon: American Mission Press, 1869.

Brass, Paul. *The Theft of an Idol: Text and Context in the Representation of Collective Violence*. Princeton, NJ: Princeton University Press, 1997.

Brotton, Jerry. *Renaissance Bazaar: From the Silk Road to Michelangelo*. Oxford: Oxford University Press, 2002.

Buragohain, Purna Kanta. *Patkair Sipare No Bochar* (*9 Years Beyond the Patkai*). Dhemaki: Purbanchal Tai Sahitya Sonha, 1993.

Burbank, Jane, and Frederick Cooper. *Empires in World History: Power and the Politics of Difference*. Princeton, NJ: Princeton University Press, 2010, Ch. 3.

Butler, Judith. 'Performativity, Precarity and Sexual Politics'. *AIBR. Revista de Antropología Iberoamericana* 4, no. 3 (2009): pp. i–xiii.

Buur, Lars. 'The Horror of the Mob: The Violence of Imagination in South Africa'. *Critique of Anthropology* 29 (2009): pp. 27–46.

Carey, Bertram Sausmarez, and Henry Newman Tuck. *The Chin Hills: A History of the People, our Dealings with them, their Customs and Manners, and a Gazetteer of their Country*, vol. 1. Rangoon: Superintendent, Government Printing, Burma, 1896.

Carey, Matthew, and Morten Axel Pedersen. 'Introduction: Infrastructures of Certainty and Doubt'. *The Cambridge Journal of Anthropology* 35, no. 2 (2017): pp. 18–29.

Cederlöf, Gunnel. *Founding an Empire on India's North Eastern Frontiers, 1790-1840: Climate, Commerce, Polity.* New Delhi: Oxford University Press, 2014.

Cederlöf, Gunnel. 'Corridors, Networks, and Pathways in India's Colonial North-east', *Fourth Jayashree Roy Memorial Lecture*, mimeo, North East India Studies Programme, Jawaharlal Nehru University, New Delhi, 2015.

Cederlöf, Gunnel. 'Seeking China's Back Door: On English Handkerchiefs and Global Local Markets in the Early Nineteenth Century'. In *Trans-Himalayan Borderlands: Livelihoods, Territorialities, Modernities*, edited by D. Smyer Yü and J. Michaud, pp. 125–144. Amsterdam: Amsterdam University Press, 2017.

Cederlöf, Gunnel. 'Tracking Routes: Imperial Competition in the Late 19th Century Burma–China Borderlands'. In *Flows and Frictions in Trans-Himalayan Spaces*. Conference held at the Royal Swedish Academy of Letters, History and Antiquities in Stockholm, Sweden, 2019, unpublished.

Cederlöf, Gunnel. *Landscapes and the Law: Environmental Politics, Regional Histories and Contests over Nature*. New Delhi: Oxford University Press, 2019 [2008].

Cederlöf, Gunnel, and K. Sivaramakrishnan, eds. *Ecological Nationalisms: Nature, Livelihoods, and Identities in South Asia*. Seattle: Washington University Press, 2014 [2005].

Chakravarti, Ranabir. 'Natural Resources and Human Settlements: Perceiving the Environment in India'. In *The World in the Year 1000*, edited by J. Heitzman and W. Schenkluhn. Lanham. MD: University Press of America, 2004.

Chakravorty, Birendra Chandra. *British Relations with the Hill Tribes of Assam since 1858*. Calcutta: K.L. Firma, 1964.

Chang, Wen-Chin. 'From a *Shiji* Episode to the Forbidden Jade Trade during the Socialist Regime in Burma'. In *Chinese Circulations: Capital, Commodities, and Networks in Southeast Asia*, edited by E. Tagliacozzo and W-C. Chang, pp. 455–480. Durham, NC and London: Duke University Press, 2011.

Chang, Wen-Chin. *Beyond Borders: Stories of Yunnanese Chinese Migrants of Burma*. Ithaca, NY and London: Cornell University Press, 2014.

Charney, Michael W. 'Literary Culture on the Burma–Manipur Frontier in the Eighteenth and Nineteenth Centuries'. *The Medieval History Journal* 14, no. 2 (2011): pp. 159–181.

Charney, Michael W. 'The Short and Long Term Roots of the Rohingya Crisis: The Eradication of a Myanmar Ethnic Group'. Cornell University Seminar and Roundtable on the Rohingya, 7 November 2017, Ithaca, NY.

Chatterji, Angana P. Thomas Blom Hansen, and Christophe Jaffrelot, eds. *Majoritarian State: How Hindu Nationalism is Changing India.* London: C. Hurst & Co., 2019.

Chatterji, N. *Monoliths and Landmarks of Mizoram.* Calcutta and Aizawl: Firma KLM and TRI, 1979.

Chatterji, Suniti Kumar. *The Place of Assam in the History and Civilisation of India.* Guwahati: University of Gauhati, 1955.

Chaube, Shibani Kinkar. *Hill politics in Northeast India.* New Delhi: Orient Longman, 1999.

Cheesman, Nick. 'How in Myanmar "National Races" Came to Surpass Citizenship and Exclude Rohingya'. *Journal of Contemporary Asia* 47, no. 3 (2017): pp. 461–483.

Chen, Wen. *The Jingtai Gazetteer of Yunnan (1455).* Kunming: Yunnan Nationalities Publishing House, 2002.

Chen, Zhaotang, and Li Qixin. *The Daoguang Gazetteer of Zhaozhou Subprefecture (1839)*, vol. 3. Taipei: The Student Book Company, 1968.

Cheng, Jinren, ed. *The Qianlong Gazetteer of Zhaozhou Subprefecture (1736).* Haikou: The Publishing House of Hainan, 2000.

Chevalier, Jean-Baptiste. *The Adventures of Jean-Baptiste Chevalier in Eastern India, 1752-1765: Historical Memoir and Journal of Travels in Assam, Bengal, and Tibet*, Translated by Caroline Dutta-Baruah and Jean Deroche. Guwahati: LBS Publications, 2008.

Chew, Emrys. *Arming the Periphery: The Arms Trade in the Indian Ocean During the Age of Global Empire.* Basingstoke: Palgrave Macmillan, 2012.

Choudhury, Deba Prosad. 'British Quest for Trade Routes from Assam to eastern Tibet: 1771–1914'. *Journal of the Royal Society for Asian Affairs* 8, no.3 (1977): pp. 180–184.

Christian, John L. 'Trans-Burma Trade Routes to China'. *Pacific Affairs* 13, no. 2 (June 1940): pp. 173–191.

Colebrooke, Henry Thomas. *Remarks on the Husbandry and Internal Commerce of Bengal.* Calcutta: n. p., 1804.

Comaroff, Jean, and John Comaroff, eds. *Law and Disorder in the Postcolony.* Chicago, IL: University of Chicago Press, 2006.

Cooper, Thomas T. *New Routes for Commerce: An Account of a Journey Made in an attempt to Penetrate Thibet from Assam to Open New Routes for Commerce.* London: Henry S. King and Co., 1873.

Coryton, J., and A. R. Margary. 'Trade Routes between British Burmah and Western China; with Extracts of Letters from Mr. Margary'. *Proceedings of the Royal Geographical Society of London* 19, no. 4 (1874–75): pp. 264–291.

Cotton, Arthur. 'On Communication between India and China by Line of the Burhampooter'. *Proceedings of the Royal Geographical Society of London* 11, no. 6 (1866): pp. 255–259.
d'Hubert, Thibaut. *In the Shade of the Golden Palace: Alaol and Middle Bengali Poetics in Arakan*. Oxford: Oxford University Press, 2018.
d'Orléans, Henri. 'A Journey from Tonkin by Tali-Fu to Assam'. *The Geographical Journal* 8, no. 6 (1896): pp. 566–582.
Dai, Yincong. 'A Disguised Defeat: The Myanmar Campaign of the Qing Dynasty'. *Modern Asian Studies* 38, no. 1 (2004): pp. 145–189.
Dalton, Edward Tuite. *Descriptive Ethnology of Bengal. By Edward Tuite Dalton. Illustrated by Lithograph Portraits Copied from Photographs. Printed for the Government of Bengal, under the Direction of the Council of the Asiatic Society of Bengal.* London: British Library, Historical Print Editions, 2011.
Daniels, Christian, and Jianxiong Ma, eds. *Transformation of Yunnan in Ming China: From the Dali Kingdom to Imperial Province.* London: Routledge, 2020.
Daniels, Christian. 'Blocking the Path of Feral Pigs with Rotten Bamboo: The Role of Upland Peoples in the Crisis of a Tay Polity in Southwest Yunnan, 1792 to 1836'. *Modern Asian Studies* 2, no. 1 (2013): pp. 133–170.
Das, Veena, and Deborah Poole, eds. *Anthropology in the Margins of the State.* Santa Fe, NM: School of American Research Advanced Seminar Series, 2004.
Dasgupta, Anindita. *Remembering Sylhet: Hindu and Muslim Voices from a Nearly Forgotten Story of India's Partition.* New Delhi: Manohar, 2014.
de Certeau, Michel. *The Practice of Everyday Life*. Berkeley, CA: University of California Press, 1984.
Deloche, Jean. *Transport and Communications in India prior to Steam Locomotion*. New Delhi: Oxford University Press, 1993.
Devi, Lakhi. *Ahom-Tribal Relations: A Political Study.* Guwahati: LBS, 1992.
'Digging deeper: Inside the Hpakant jade mine'. *Myanmar Times*, 8 June 2016.
Digby, Simon. *War-Horse* Chang, Wen-Chin. 'From a *Shiji* Episode to the Forbidden Jade Trade during the Socialist Regime in Burma'. In *Chinese Circulations: Capital, Commodities, and Networks in Southeast Asia*, edited by E. Tagliacozzo and W-C. Chang, pp. 455–480. Durham, NC and London: Duke University Press, 2011.
Digby, Simon. *War-Horse and Elephant in the Delhi Sultanate: A Study of Military Supplies.* Oxford: Orient Monographs, 1971.
Dikötter, Frank. *The Discourse of Race in Modern China*. Hong Kong: Hong Kong University Press, 1992.
'Dimapur Lynching: It was "Consensual Sex" not Rape, says Nagaland Govt Report'. *Indian Express*, 12 March 2015.
Douglas, Mary. *Purity and Danger: An Analysis of Concepts of Pollution and Taboo.* London: Routledge and Kegan Paul, 1966.

'Drive to Flush Out Migrants'. *The Telegraph*, 11 February 2015.

Dutta, Keshab Narayan. *Landmarks of the Freedom Struggle.* Guwahati: LBS, 1958.

Dutta, Keshab Narayan. *A Hand Book to the Old Records of the Assam Secretariat.* Shillong: Assam Government Press, 1959.

Dutta, Nilim. 'Dimapur Lynching: They called him a Bangladeshi and a Rapist just to Kill him' (2015), accessed 2 December 2021, https://www.dailyo.in/politics/dimapur-lynching-nagaland-sharif-uddin-khan-bangladeshi-immigrants/story/1/2543.html.

Dzüvichü, Lipokmar. 'The Colonial State and "Illegal" Arms Trade along the North East Frontier of British India, 1860s–1900s'. In *Landscape, Culture and Belonging: Writing the History of Northeast India*, edited by N. Bhattacharya and J. L. K. Pachuau. Cambridge: Cambridge University Press, 2019.

Eaton, Richard M. *The Rise of Islam and the Bengal Frontier, 1204–1760*. Berkeley and Los Angeles, CA: University of California Press, 1994.

Egreteau, Renaud. 'The Burmese Jade Trail: Transnational Networks, China, and the (Relative) Impact of International Sanctions on Myanmar's Gem Industry'. In *Myanmar's Transition: Openings, Obstacles and Opportunities*, edited by N. Cheesman, M. Skidmore, and T. Wilson, pp. 89–116. Singapore: Institute of Southeast Asian Studies, 2012.

Elverskog, Johan. *Buddhism & Islam on the Silk Road*. Philadelphia, PA: University of Pennsylvania Press, 2010.

Elvin, Mark. *The Retreat of the Elephants: An Environmental History of China*. New Haven, CN and London: Yale University Press, 2004.

Elwin, Verrier. *Nagaland.* Shillong: The Research Department Secretariat, 1961.

Endle, Sidney. *The Kacharis*. London: Macmillan, 1911.

Fang, Guoyu. *Research on Historical Geography of Southwest China*. Beijing: Zhonghua Book Company, 1987.

Fisher, Thomas. 'Timber Trade in Cachar'. *Journal of the Asiatic Society of Bengal* 1 (1832): pp. 305–306.

'Five Muslim Groups Urged to accept NVC'. *BBC Burmese*, accessed 31 July 2020, https://www.bbc.com/burmese/42256537.

Frey, Robert, and Bertil Lintner. 'Jade in Burma. The Major Jadeite Source'. In *Jade*, edited by Roger Keverne, pp. 267–272. New York, NY: Van Nostrand Reinhold, 1991.

Frydenlund, Shae A. 'Motherhood, home, and the political economy of Rohingya women's labor'. In *Unraveling Myanmar's Transition*, edited by Pavin Chachavalpongpun, Elliott Prasse-Freeman, and Patrick Strefford, pp. 236–241. Singapore: NUS Press, 2020.

Frydenlund, Shae A., and S. Lei. 'Double Exclusions, Double Shifts: Muslim Women's Labor Subjectivities in Yangon, Myanmar'. *Independent Journal of Burmese Scholarship* (forthcoming).

Fur, Gunlög. 'Concurrences as a Methodology for Discerning Concurrent Histories'. In *Concurrent Imaginaries, Postcolonial Worlds: Toward Revised Histories*, edited by D. Brydon, P. Forsgren, and G. M. Fur, pp. 33–57. Leiden and Boston, MA: Brill Rodopi, 2017.

Fürer-Haimendorf, Christoph, von. *The Naked Nagas*. London: Methuen and Co., 1933.

Gait, Edward. *A History of Assam*. Calcutta: Thacker and Sprink, 1906.

Galache, Carlos Sardiña, and Antolin Avezuela. 'Citizenship for a few, rights for none'. *Equal Times*, 24 July 2017, as quoted in https://www.aljazeera.com/indepth/features/2017/08/rohingya-muslims-170831065142812.html, accessed 30 July 2020.

Gallagher, John, and Ronald Robinson. 'The Imperialism of Free Trade'. *Economic History Review* 6, no. 1 (1953): pp. 1–15.

Ganguly, Jayanta Bhusan. 'The Modus Operandi of Trade between the Eastern Himalayan Subregion and Assam in the Pre-Colonial and Colonial Periods'. In *Border Trade: North East India and Neighbouring Countries*, edited by G. Das and R.K. Purakayastha. New Delhi: Akansha Publishing House, 2000.

Gao, David. 'Railways and the issue of Inner Line Permit in Arunachal Pradesh: Can the two function together?'. *Economic and Political Weekly* 50, no. 8 (2015), https://www.epw.in/journal/2015/8/reports-states-web-exclusives/railways-and-issue-inner-line-permit-arunachal-pradesh, accessed 20 October 2019..

Geschiere, Peter. *The Perils of Belonging: Autochthony, Citizenship, and Exclusion in Africa and Europe*. Chicago, IL: Chicago University Press, 2009.

Giersch, Charles Patterson. *Asian Borderlands: The Transformation of Qing China's Yunnan Frontier*. Cambridge, MA and London: Harvard University Press, 2006.

Giersch, Charles Patterson. 'Cotton, Copper and Caravans: Trade and the Transformation of Southwest China'. In *Chinese Circulations: Capital, Commodities, and Networks in Southeast Asia*, edited by E. Tagliacozzo and W-C. Chang, pp. 37–61. Durham, NC and London: Duke University Press, 2011.

Girard, René. *The Scapegoat*. Baltimore, MD: Johns Hopkins University Press, 1989.

Global Witness. *Jade: Myanmar's "big state secret"*. London: Global Witness Limited, 2015.

Goldstein, Daniel M. '"In our own Hands": Lynching, Justice, and the Law in Bolivia'. *American Ethnologist* 30 no. 1 (2003): pp. 22–43.

Goodenough, F. A. 'Letter to Major-General Sir Andrew Scott Waugh, on Routes between Upper Assam and Western China'. *Proceedings of the Royal Geographical Society of London* 12, no. 5 (1867–68): pp. 334–336.

Goswami, Manu. *Producing India: From Colonial Economy to National Space*. Chicago, IL: University of Chicago Press, 2004.

Goswami, Shrutidev. *Aspects of Revenue Administration in Assam 1826-1874*. New Delhi: Mittal, 1987.

Graeber, David. *Toward an Anthropological Theory of Value*. New York: Palgrave, 2001.

Granoff, Phyllis. 'Illuminating the Formless, God, King, and Devotion in Assamese Illustrated Manuscript'. In *Bhakti and Power: Debating India's Religion of the Heart*, edited by C. L. Novetzke, S. Sharma, and J. Stratton Hawley, p. 131. Seattle, WA: University of Washington Press, 2019.

Gribble, R. H. *Out of the Burma Night: Being the Story of a Fantastic Journey through the Wilderness of the Hukawng Valley and the Forest Clad Mountains of the Naga Tribes People at the Time of the Japanese Invasion of Burma.* Calcutta: Thacker Spink & Co., 1944.

Grierson, George Abraham. *Linguistic Survey of India, Indo-Aryan Family. Eastern Group. Specimens of the Bengali and Assamese Language*, vol. 5. Calcutta: Government Printing, 1903.

Griffith, William. 'Report on the Caoutchouc Tree of Assam made at the Request of Captain Jenkins Agent to the Governor General'. *Journal of the Asiatic Society of Bengal* 7 (1838): pp. 132–142.

Guha, Amalendu. 'A Big Push Without a Take-off: A Case-Study of Assam, 1871-1901'. *The Indian Economic and Social History Review* 5, no. 3 (1968): pp. 199–221.

Guha, Amalendu. 'The Medieval Economy of Assam'. In *The Cambridge Economic History of India*, edited by D. Kumar and T. Raychaudhuri. Cambridge: CUP, 1982.

Guha, Amalendu. 'The Ahom Political System: An Enquiry into State Formation in Medieval Assam, 1228-1800'. In *Tribal Polities and State Systems in Pre-Colonial East and Northeastern India*, edited by Surajit Sinha. Calcutta: KP Bagchi and Co., 1987.

Guha, Sumit, and Kenneth Pomeranz. 'Life and Energy'. In *What China and India Once Were: The Pasts That May Shape the Global Future*, edited by B. Elman and S. Pollock. New Delhi: Penguin, 2018.

Guite, Jangkhomang. *Against State, against History: Freedom, Resistance, and Statelessness in Upland Northeast India*. Delhi: Oxford University Press, 2019.

Gurdon, Philip Richard Thornhagh (Major). *The Khasis*. London: David Nutt, 1907.

Guyot-Réchard, Bérénice. *Shadow States: India, China and the Himalayas, 1910-1962*. Cambridge: CUP, 2017.

Hall, Daniel George Edward. *A History of South-East Asia*. London and New York: Macmillan and St. Martin's Press, 1968.

Hamilton, Angus. *In Abor Jungles: Being an Account of the Abor Expedition, the Mishmi Mission and the Miri Mission.* London: Eveleigh Nash, 1912.

Hamilton, Francis. *An Account of Assam [First Compiled in 1807-14] Edited by Surya Kumar Bhuyan.* Guwahati: Department of Historical and Antiquarian Studies, 1940.

Hansen, Thomas Blom, and Finn Stepputat, eds. *Sovereign Bodies: Citizens, Migrants, and States in the Postcolonial World.* Princeton, NJ: Princeton University Press, 2005.

Hansen, Valeria. *The Silk Road: A New History.* New York, NY: Oxford University Press, 2012.

Haokip, Thongkholal. *India's Look East Policy and the Northeast.* New Delhi: Sage, 2015.

Harris, Tina. *Geographical Diversions: Tibetan Trade, Global Transactions.* Athens, GA: University of Georgia Press, 2013.

Harvey, David. *The Condition of Postmodernity: An Enquiry into the Origins of Cultural Change.* Cambridge, MA and Oxford: Blackwell, 1990.

Harvey, G. E. *History of Burma.* Calcutta: AES, 2000.

Hasnu, Santosh. 'A History of Transport in Northeast India, 1824-1945', PhD diss., University of Delhi, unpublished.

Hay, John Ogilvy. *A Map Shewing the Various Routes Connecting China with India and Europe through Burmah and Developing the Trade of Eastern Bengal, Burmah & China.* London: Edward Stanford, 1875.

Hazarika, Nilam. 'A Study into the Ahom Military System in Medieval Assam'. *International Journal of Humanities and Social Science Invention ISSN* 4, no. 6 (2015): pp. 17–22.

Hechter, Michael. *Internal Colonialism: The Celtic Fringe in British National Development, 1536-1966.* International Library of Sociology. London: Routledge and Kegan Paul, 1975.

Heitzman, James. *Gifts of Power: Lordship in an Early Indian State.* New Delhi: Oxford University Press, 1997.

Heng, Geraldine, and Janadas Devan. 'State Fatherhood: The Politics of Nationalism, Sexuality, and Race in Singapore'. In *Nationalism and Sexualities*, edited by A. Parker, M. Russo, D. Sommer, and P. Yeager. New York, NY: Routledge. 1992.

Henniker, F. C. *Gold and Silver Wares of Assam: A Monograph.* Shillong: Assam Secretariat Printing Office, 1905.

Henry, Augustine. 'Chinese Names of Plants'. *Journal of the China Branch of the Royal Asiatic Society* 22–23 (1887): p. 245.

Herman, John E. *Amid the Clouds and Mist: China's Colonization of Guizhou, 1200-1700.* Cambridge, MA and London: Harvard University Asia Center, Harvard University Press. 2007.

Holdich, Thomas Hungerford. 'The North-Eastern Frontier of India'. *Journal of the Royal Society of Arts* 60, no. 3092 (1912): p. 387.

Horam, Mashangthei. *Social and Cultural Life of Nagas (The Tangkhul Nagas).* New Delhi: BR Publishing Corporation, 1977.

Horam, Mashangthei. *The Naga Insurgency: The Last Thirty Years.* New Delhi: Cosmo Publications, 1988.

Hostetler, Laura. *Qing Colonial Enterprise: Ethnography and Cartography in Early Modern China.* Chicago, IL and London: University of Chicago Press, 2001.

House of Commons Parliamentary Papers, Reports on Trade Routes and Fairs on the Northern Frontiers of India, (London, Her Majesty's Stationery Office, 1874).
Htin, Kyaw Minn. 'Where Mandalas Overlap: Histories, identities and Fates of the People from Arakan and South-Eastern Bangladesh', PhD diss., National University of Singapore, 2017.
Hucker, Charles O. *A Dictionary of Official Titles in Imperial China*. Taipei: Reprinted by Southern Materials Center INC, 1985.
Hughes, Richard W. 'Burma's Jade Mines. An Annotated Occidental History'. *Lotus Gemology* (1999), accessed 2 December 2021, http://www.lotusgemology.com/index.php/library/articles/283-burma-s-jade-mines-an-annotated-occidental-history.
Humphrey, Caroline. 'Introduction: Trusting and Mistrusting Across Borders'. In *Trust and Mistrust in the Economies of the China-Russia Borderlands*, edited by Caroline Humphrey, pp. 9–37. Amsterdam: Amsterdam University Press, 2018.
Hunter, William Wilson. *A Statistical Account of Assam*, vol. 1. London: Trubner, 1879.
Hussain, Bonojit. 'Nation's Honour: The Dimapur Lynching and "Illegal Bangladeshi Immigrants"' (2015), accessed 2 December 2021, https://scroll.in/article/712715/nations-honour-the-dimapur-lynching-and-illegal-bangladeshi-immigrants.
Hussain, Syed Ejaz. 'Silver Flow and Horse Supply to Sultanate Bengal with Special Reference to Trans-Himalayan Trade (13th-16th Centuries)'. *Journal of the Economic and Social History of the Orient* 56, no. 2 (2013): pp. 264–308.
Hutton, John Henry. *The Angami Nagas*. London: Macmillan, 1921.
Hutton, John Henry. 'Naga Chank Ornaments of South Indian Affinities'. *Man* 26 (1926): pp. 222–224.
Ibrahim, Azeem. *The Rohingyas: Inside Myanmar's Hidden Genocide.* London: C. Hurst and Company (Publishers) Limited, 2016.
Interview with U Kyaw Min, Yangon, 8 November 2019.
Iqbal, Iftekhar. 'The Space between Nation and Empire: The Making and Unmaking of Eastern Bengal and Assam Province 1905-1911'. *Journal of Asian Studies* 74, no. 1 (2015): p. 74.
Iralu, Kaka. *The Naga Saga: A Historical Account of the Sixty-Two Years Indo-Naga War and the Story of Those Who were Never Allowed to Tell it.* Kohima: K. Iralu, 2002.
'It is finally clear'. *News Eleven*, 26 February 2018, accessed 31 July 2020, https://news-eleven.com/features/40345.
Jenkins, Henry Lionel. 'Note on the Burmese Route from Assam to the Hookoong Valley'. In *Selection of Papers Regarding the Hill Tracts between Assam and Burmah and the Upper Brahmaputra*, pp. 251–254. Calcutta: Bengal Secretariat Press, 1869–70.
Jhala, Angma Dey. *An Endangered History: Indigeneity, Religion, and Politics on the Borders of India, Burma, and Bangladesh*. New Delhi: Oxford University Press, 2019.

Jones, Lee. 'Understanding Myanmar's Ceasefires: Geopolitics, Political Economy and State-Building'. In *War and Peace in the Borderlands of Myanmar: The Kachin Ceasefire, 1994-2011*, edited by Mandy Sadan, pp. 95-113. Copenhagen: Nias Press, 2016.

Kachin Development Network Group (KDNG). *Blood Amber: Military Resource Grab Clears out Indigenous Peoples in Kachin State's Hugawng Valley* (2019), accessed 2 December 2021, https://kdng.org/wp-content/uploads/2019/08/BloodAmber_English.pdf.

Kakati, Banikanta. *Assamese: Its Formation and Development: A Scientific Treatise on the History and Philology of the Assamese Language.* Guwahati: LBS, 1987 [1941].

Kanungo, Alok Kumar. 'Naga Ornaments and the Indian Ocean'. *Bulletin of the Indo-Pacific Prehistory Association* 26 (2006): pp. 154–162.

Kar, Bodhisattva. 'Historia Elastica: A Note on the Rubber Hunt in the North-Eastern Frontier of British India'. *Indian Historical Review* 36, no. 1 (June 2009): pp. 131–150.

Kar, Bodhisattva. 'When was the Postcolonial? A History of Policing Impossible Lines'. In *Beyond Counter-Insurgency: Breaking the Impasse in Northeast India*, edited by Sanjib Baruah, pp. 39–80. Oxford: Oxford University Press, 2009.

Kar, Bodhisattva. 'Nomadic Capital and Speculative Tribes: A Culture of Contracts in the Northeastern Frontier of British India'. *The Indian Economic & Social History Review* 53, no. 1 (2016): pp. 41–67.

Karlsson, Bengt G. *Unruly Hills: A Political Ecology of India's Northeast.* New York, NY and Oxford: Berghahn, 2011.

Kew Royal Botanic Gardens. *Official Guide to the Museums of Economic Botany, issue 1, Dicotyledons and Gymnosperms.* London: Her Majesty's Stationery Office, 1883.

Kiik, Laur. 'Conspiracy, God's Plan and National Emergency: Kachin Popular Analyses of the Ceasefire Era and its Resource Grabs'. In *War and Peace in the Borderlands of Myanmar: The Kachin Ceasefire, 1994-2011*, edited by Mandy Sadan, pp. 205–235. Copenhagen: Nias Press, 2016.

Kikon, Dolly. 'The Predicament of Justice: 50 Years of Armed Forces Special Powers Act'. *Journal of Contemporary South Asia* 17, no. 3 (2009): pp. 271–282.

Kikon, Dolly. 'The City of Sorrow: Revisiting the 2015 Dimapur Lynching' (2015), accessed 2 December 2021, https://scroll.in/article/729595/the-city-of-sorrow-revisiting-the-2015-dimapur-lynching.

Kikon, Dolly. 'What is Unique about Naga History'. *Economic and Political Weekly* 50, no. 35 (2015): pp. 10–13.

Kikon, Dolly. *Life and Dignity: Women's Testimonies of Sexual Violence in Dimapur (Nagaland).* Guwahati: North-Eastern Social Research Centre, 2016.

Kikon, Dolly. 'Difficult Loves: Purity and Hardship'. In *Northeast India: A Place of Relations*, edited by Y. Saikia and A. R. Biashya, pp. 225–243. Cambridge: Cambridge University Press, 2017.

Kikon, Dolly. 'From the Heart to the Plate: Debates about Dog Meat in Dimapur'. *International Institute for Asian Studies Newsletter* 77 (2017): pp. 38–39.

Kyaw, Nyi Nyi. 'Islamophobia in Buddhist Myanmar: The 969 Movement and Anti-Muslim Violence'. In *Islam and The State in Myanmar*, edited by Melissa Crouch, pp. 183–210. Oxford: Oxford University Press, 2016.

Laclau, Ernesto. *On Populist Reason*. London: Verso Books, 2005.

Laldinpuii, K. 'A Memorial Stone Platform of Lapuithanga Sailo Chief'. In *A Modern History of Mizoram*, edited by Sangkima. Guwahati: Spectrum Publications, 2004.

'Lawless Christian Intolerance'. *Eastern Mirror*, 9 March 2015.

Leach, Edmund. *Political Systems of Highland Burma: A Study of Kachin Social Structure*. London: The Athlone Press, 1970 [1954].

Lee, James. 'Food Supply and Population Growth in Southwest China, 1250-1850'. *Journal of Asian Studies* 41, no. 4 (1982): pp. 711–746.

Lehman, F. K. *The Structure of Chin Society: A Tribal People of Burma Adapted to a Non-Western Civilization*. Urbana, IL: University of Illinois Press, 1963.

Levy, Adrian, and Cathy Scott-Clark. *The Stone of Heaven: The Secret History of Imperial Green Jade*. London: Weidenfeld and Nicholson, 2001.

Lewin, Thomas Herbert. *The Hill Tracts of Chittagong and the Dwellers Therein; With Comparative Vocabularies of the Hill Dialects*. Calcutta: Bengal Printing Company, 1869.

Li, Fuyi. *The Letters of Fulu, Taipei*. Taipei: Furen Book Company, 2005.

Li, Tania M. 'Ethnic Cleansing, Recursive Knowledge, and the Dilemmas of Sedentarism'. *International Social Science Journal* 54, no. 173 (2002): pp. 361–371.

Lieberman, Victor B. *Strange Parallels: Southeast Asia in Global Context, c. 800–1830*. Cambridge: Cambridge University Press, 2003.

Lindsay, Robert. 'Anecdotes of an Indian Life'. In *Lives of the Lindsays or A Memoir of the Houses of Crawford and Balcarres*, edited by Lord Lindsay. London: John Murray, 1849.

Liu, Wenzheng, ed. *The Tianqi Gazetteer of Yunnan (1625)*. Kunming: Yunnan Education Publishing, 1991.

Long, Chunlin, and Ji Huang. 'Traditional Cultivation of Coptis Teeta and its Values in Biodiversity Conservation'. *Biodiversity Science* 14, no. 1 (2006): pp. 79–86.

Long, Chunlin, and Ji Huang. 'Coptis Teeta-Based Agroforestry System and Its Conservation Potential: A Case Study from Northwest Yunnan'. *AMBIO: A Journal of the Human Environment* 36, no. 4 (2007): pp. 343–349.

Lotha, Abraham. 'Articulating Naga Nationalism'. PhD diss., City University of New York, 2009.

Lu, Jiarui, ed. *Gazetteer of Ethnic Minorities and Religions in Jianchuan County*. Kunming: Yunnan Nationalities Publishing House, 2003.

Ludden, David. *Making India Hindu: Religion, Community, and the Politics of Democracy in India*. New Delhi: Oxford University Press, 1996.

Ludden, David. *An Agrarian History of South Asia*. Cambridge: Cambridge University Press, 1999.

Ludden, David. *Reading Subaltern Studies: Critical History, Contested Meaning and the Globalization of South Asia*. London: Anthem, 2002.

Ludden, David. 'Maps in the Mind and the Mobility of Asia'. *Journal of Asian Studies* 62, no. 4 (2003): pp. 1057–1078.

Ludden, David. 'The First Boundary of Bangladesh on Sylhet's Northern Frontiers'. *Journal of the Asiatic Society of Bangladesh* 48, no. 1 (June 2003): pp. 1–54.

Ludden, David. 'The Process of Empire: Frontiers and Borderlands'. In *Tributary Empires in Global History*, edited by P. Fibiger Bang and C. A. Bayly, pp. 132–150. London: Palgrave Macmillan, 2011.

Ludden, David. 'Imperial Modernity: History and Global Inequity in Rising Asia'. *Third World Quarterly* 33, no. 4 (2012): pp. 581–601.

Ludden, David. *India and South Asia: A Short History*. 2nd ed. London: OneWorld, 2014.

Ludden, David. 'The Decline of Buddhism, Revisited: The Rise of Hindu Territorial Hegemony, c.950–1250'. In *Clio and Her Descendants: Essays for Kesavan Veluthat*, edited by Manu V. Mahadevan, pp. 132–162. New Delhi: Primus Books, 2018.

Ludden, David. 'Cowry Country: Mobile Space and Imperial Territory'. In *Asia Inside Out. Volume III, Itinerant People*, edited by E. Tagliacozza, H. F. Siu, and P. Perdue, pp. 75–100. Cambridge, MA: Harvard University Press, 2019.

Ludden, David. 'India's Spatial History in the Brahmaputra-Meghna River Basin'. In *Landscape, Culture, and Belonging: Writing the History of Northeast India*, edited by Neeladri Bhattacharya and Joy L.K. Pachuau, pp. 23–37. Cambridge: Cambridge University Press, 2019.

M'Cosh, John. *Topography of Assam*. Calcutta: Bengal Military Orphan Press, 1837.

Ma, Cunzhao, ed. 'Inscription of Hongshan village'. In *Collection of Historical Inscriptions in Fengyi*. Hong Kong: South China Research Center, 2013.

Ma, Cunzhao, ed. 'Inscription on Reconstruction of Flown Monastery'. In *Collection of Historical Inscriptions in Fengyi*. Hong Kong: South China Research Center, 2013.

Ma, Cunzhao, ed. 'Inscription on the Construction of Zhaozhou City'. In *Collection of Historical Inscriptions in Fengyi*. Hong Kong: South China Research Center, 2013.

Ma, Cunzhao, ed. 'Inscription on the Corvee Assignments (1662)'. In *Collection of Historical Inscriptions in Fengyi*. Hong Kong: South China Research Center, 2013.

Ma, Cunzhao, ed. 'Inscription on the Temple Property of Wenchang Temple (1695)'. In *Collection of Historical Inscriptions in Fengyi*. Hong Kong: South China Research Center, 2013.

Ma, Jianxiong. 'Marriage and Land Property: Bilateral Non-Lineal Kinship and Communal Authority of the Lahu on the Southwest Yunnan Frontier, China'. *South East Asia Research* 19, no. 3 (2011): pp. 495–536.

Ma, Jianxiong. 'Shaping of the Yunnan-Burma Frontier by Secret Societies since the End of the 17th Century'. *Moussons N°* 17, no. 1 (2011): pp. 65–84.

Ma, Jianxiong. *The Lahu Minority in Southwest China: A Response to Ethnic Marginalization on the Frontier.* Abingdon, Oxon: Routledge, 2013.

Ma, Jianxiong. 'The Rise of Gentry Power on the China-Burma Frontier since the 1870s: The Case of the Peng Family in Mianning, Southwest Yunnan'. *International Journal of Asian Studies* 11, no. 1 (2014): pp. 25–51.

Ma, Jianxiong. 'The Zhaozhou Bazi Society in Yunnan: Historical Process in the Bazi Basin Environmental System During the Ming Period (1368-1643)'. In *Environmental History in East Asia: Interdisciplinary Perspectives*, edited by Ts'ui-jung Liu, pp. 131–155. Abingdon and New York, NY: Routledge, 2014.

Ma, Jianxiong. 'Salt and Revenue in the Frontier Formation: State Mobilized Ethnic Politics in Yunnan-Burma Borderland since the 1720s'. *Modern Asian Studies* 48, no. 6 (2014): pp. 1637–1669.

Ma, Jianxiong. 'State and Reconstruction of the Southwest Frontier: Roads, Resources and Ritual Interpretations'. In *The Bazi Society of the Yun-Gui Plateau: Transportation Route, Resources and Ritual Interpretation*, edited by Z. Min and T-S. Liu, pp. 1–19. Kunming: Yunnan University Press, 2018.

Ma, Jianxiong. 'The Dowry Land System and Chieftains of Shan-Dai Borderlands between Yunnan and Burmese Kingdoms from the Ming to the Qing Dynasties: The Construction of a Decentralized Institution in the Frontier'. In *The India-China Corridor workshop 'Between Yunnan and Bengal: Process geographies in the making of Modern Asia'.* Yunnan Minzu University, Kunming, China, 17–19 April 2018.

Ma, Jianxiong. 'Local Communities, Village Temples and the Reconstruction of Ethnic Groups in Western Yunnan, 14th to 17th Centuries'. In *Transformation of Yunnan in Ming China: From the Dali Kingdom to Imperial Province*, edited by C. A. Daniels and J. Ma, pp. 43–75. London: Routledge, 2020.

Ma, Jianxiong. 'Constructing Native Chieftains as Imperial Frontier Institution: Endogamy and Dowry Land Exchange among the Shan-Dai Chieftains in Yunnan–Burma Borderland since the Thirteenth Century'. In *Yunnan–Burma–Bengal Corridor Geographies Protean Edging of Habitats and Empires*, edited by D. S. Yü and K. Dean, Ch. 8. New Delhi: Routledge, 2022.

Ma, Jianxiong., and Cunzhao Ma. 'The Mule Caravans of Western Yunnan: An Oral History of the Muleteers of Zhaozhou'. *Transfers* 4, no. 3 (2014): pp. 24–42.

Ma, Jianxiong, and Cunzhao Ma. 'The Mule Caravans as Cross-Border Networks: Local Bands and Their Stretch on the Frontier between Yunnan and Burma'. In

Myanmar's Mountain and Maritime Borderscapes: Local Practices, Boundary-Making and Figured Worlds, edited by Su-Ann Oh, pp. 237–257. Singapore: ISEAS Publishing, 2016.

Mack, Rosamond E. *Bazaar to Piazza: Islamic Trade and Italian Art, 1300–1600*. Berkeley, CA: University of California Press, 2002.

Mackenzie, Alexander. *History of the Relations of the Government with the Hill Tribes of the North-east Frontier of India.* Calcutta: Home Department Press, 1884.

Mackenzie, Alexander. *The Northeast Frontier of India,* [first published as *History of the Relations of the Government with the Hill Tribes of the Northeast Frontier of India,* 1884]. Delhi: Mittal, 1979.

Maksic, Adis. *Ethnic Mobilization, Violence, and the Politics of Affect: The Serb Democratic Party and the Bosnian War.* Cham: Palgrave Macmillan, 2017.

Malekandathil, Pius. 'Circulation of People and Patterns of Maritime Migrations in the Indian Ocean during the Pre-modern Period'. In *Migrations in Medieval and Early Colonial India*, edited by Vijaya Ramaswamy, pp. 259–286. London: Routledge, 2016.

Malsawmliana. 'Megalithic Monuments of the Mizos'. *Historical Journal Mizoram* VI (2005): pp. 52–62.

Mamdani, Mahmood. *When Victims Become Killers: Colonialism, Nativism, and the Genocide in Rwanda.* Princeton, NJ: Princeton University Press, 2001.

Mangrai, Sao Saimong. 'The Shan States and the British Annexation'. Ithaca, NY: Cornell University, 1965.

Manjapra, Kris. *Colonialism in Global Perspective.* Cambridge: Cambridge University Press, 2020.

Markovits, Claude, Jacques Pouchepadass, and Sanjay Subrahmanyam. *Society and Circulation: Mobile People and Itinerant Cultures in South Asia, 1750-1950.* London: Anthem Press, 2006.

Marsden, Magnus. 'Actually Existing Silk Roads'. *Journal of Eurasian Studies* 8, no. 1 (2017).

Marx, Karl. *Grundrisse: Foundations of the Critique of Political Economy.* London: Penguin, 1973.

Maung, Zarni. 'The official evidence of the Rohingya ethnic ID and citizenship which the Burmese ethno- and genocidists don't want you to see'. *Rohingya Blogger* http://www.rohingyablogger.com/2013/05/the-official-evidence-of-rohingya.html, accessed 31 July 2020.

Maxwell Hill, Ann. *Merchants and Migrants: Yunnanese Chinese in Southeast Asia.* New Haven, CN: Yale University Southeast Asia Studies, 1998.

McAuliffe, Erin L. 'Caste and the quest for racial hierarchy in British Burma: An analysis of census classifications from 1872–1931', MA diss., University of Washington, 2017.

McCosh, John. 'Account of the Mountain Tribes on the Extreme N.E. Frontier of Bengal'. *Journal of the Asiatic Society of Bengal* 203 (1836).

McDuie-Ra, Duncan. 'Anti-Development or Identity Crisis? Misreading civil society in Meghalaya, India'. *Asian Ethnicity* 8, no. 1 (2007): pp. 43–59.

McDuie-Ra, Duncan. 'Adjacent Identities in Northeast India'. *Asian Ethnicity* 17, no. 3 (2016): pp. 400–413.

Medhi, Kaliram. *Assamese Grammar and Origin of the Assamese Language.* Guwahati: Publication Board, Assam, 1988.

Michaud, Jean. 'What's (Written) History For? On James Scott's Zomia, Especially Chapter 6½'. *Anthropology Today* 33, no. 1 (2017): pp. 6–10.

Middleton, Townsend. 'Anxious Belongings: Anxiety and the Politics of Belonging in Subnationalist Darjeeling'. *American Anthropologist* 115, no. 4 (2013): pp. 608–621.

Milko, Victoria. 'Genocide card: Myanmar Rohingya verification scheme condemned'. *Aljazeera*, 2 September 2019, accessed 30 July 2020, https://www.aljazeera.com/news/2019/09/genocide-card-myanmar-rohingya-verification-scheme-condemned-190903012922259.html.

Miller, Arthur, and Michael Davis. *Intellectual Property: Patents, Trademarks and Copyright.* Minnesota, MN: West Publishing Company, 1983.

Mills, John P. 'The Mishmis of the Lohit Valley, Assam'. *The Journal of the Royal Anthropological Institute of Great Britain and Ireland* 82, no. 1 (1952): pp. 1–12.

Mitra, Anirban, and Debraj Ray. 'Implications of an Economic Theory of Conflict: Hindu-Muslim Violence in India'. *Journal of Political Economy* 122, no. 4 (2014): pp. 719–765.

Møller, Henrik Kloppenborg. 'Becoming Real Businessmen? Negotiating Huangniu Identities at a Fake Commodity Market in Shanghai', MSc. diss., University of Copenhagen, 2010.

Møller, Henrik Kloppenborg. 'The Potentials of *Feicui*: Indeterminacy and Determination in Human-Jade Interactions in South-West China'. In *Emptiness and Fullness: Ethnographies of Lack and Desire in Contemporary China*, edited by S. Bregnbæk and M. Bunkenborg, pp. 120–139. New York, NY and Oxford: Berghahn Books, 2017.

Møller, Henrik Kloppenborg. 'Spectral Jade: Materiality, Conceptualisation, and Value in the Myanmar-China Jadeite Trade', PhD diss., Lund University, 2019.

Møller, Henrik Kloppenborg. 'Borderlines, Livelihood, and Ethnicity in the Yunnan–Myanmar Borderlands'. In *Yunnan–Burma–Bengal Corridor Geographies: Protean Edging of Habitats and Empires*, edited by Dan Smyer Yü and Karin Dean, pp. 122-142. New Delhi and London: Routledge India, 2021.

Morgan, Llewelyn. *The Buddhas of Bamiyan.* Cambridge, MA: Harvard University Press, 2012.

'Muslim students from Sittwe donated money to Rakhine war victims'. *Narijara News*, 23 July 2020, https://burmese.narinjara.com/news/detail/5f194fe4700aa70862a0935c, accessed 31 July 2020.

'Muslims asked to perform good deeds', *DVB*, 9 April 2013, accessed 16 July 2020, http://burmese.dvb.no/archives/38299.

Myat Mon, Su. 'Muslim Woman from Rakhine Jailed for Attempted Yangon Travel', *Frontier*, 2 June 2018, accessed 26 October 2019, https://frontiermyanmar.net/en/muslim-woman-in-rakhine-jailed-for-attempted-yangon-travel.

Myint-U, Thant. *The Making of Modern Burma*. New York, NY: Cambridge University Press, 2001.

'Naga Council Dimapur and Naga Women Hoho Dimapur Condemn'. *Eastern Mirror*, 3 March 2015.

'Naga Mob was Wrong'. *Morung Express*, 19 March 2015.

Naylor, Thomas. 'The Underworld of Gemstones: Part I – Under the Rainbow'. *Crime, Law & Social Change* 53, no. 2 (2010): pp. 131–158.

Neelis, Jason Emmanuel. *Early Buddhist Transmission and Trade Networks: Mobility and Exchange Within and Beyond the Northwestern Borderlands of South Asia*. Leiden: Brill, 2010.

Newell, Stephanie. *West African Literatures: Ways of Reading.* Oxford: Oxford University Press, 2006.

Newell, Stephanie. 'Constructions of Nigerian Women in Popular Literatures by Men'. *African Languages and Cultures* 9, no. 2 (1996): pp. 169–188.

Ni, Tui. *The Chronicle of Yunnan (1846)*. Kunming: Yunnan University Press, 1992.

Notar, Beth E. 'Authenticity Anxiety and Counterfeit Confidence: Outsourcing Souvenirs, Changing Money, and Narrating Value in Reform-Era China'. *Modern China* 32, no. 1 (2006): pp. 64–98.

'NSCN-IM on Jamir's Article'. *Morung Express*, 12 January 2015.

'NSF Against IBIs: Rally to be Held Today'. *Eastern Mirror*, 4 March 2015.

Nyiri, Pál. 'Enclaves of Improvement: Sovereignty and Developmentalism in the Special Zones of the China-Lao Borderlands'. *Comparative Studies in Society and History* 54, no. 3 (2012): pp. 533–562.

Ong, Aihwa. *Flexible Citizenship: The Cultural Logics of Transnationality.* Durham, NC and London: Duke University Press, 1999.

Openjadedata.org. 'Even with New Data, Valuing Myanmar's Jade Industry Remains a Challenge', accessed 2 December 2021, https://openjadedata.org/stories/how_much_jade_worth.html.

Osburg, John. 'Consuming belief. Luxury, authenticity, and Chinese patronage of Tibetan Buddhism in contemporary China'. *HAU: Journal of Ethnographic Theory* 10, no. 1 (2020): pp. 69–84.

Pachuau, Joy L. K. *Being Mizo: Identity and Belonging in Northeast India*. New Delhi: Oxford University Press, 2014.

Pachau, Joy L. K., and Willem van Schendel. *The camera as witness: A social history of Mizoram, Northeast India*. London: Routledge, 2015.

Pachuau, Joy L. K., and Willem van Schendel. 'Borderland Histories, Northeastern India: An Introduction'. *Studies in History* 32, no. 1 (2016): pp. 1–4.

Pachuau, Lalsangkima. '"Assistants" or "Leaders"? The Contributions of Early Native Christian Converts in Northeast India'. In *Christianity in Indian History*, edited by P. Malekandathil, J. L. K. Pachuau, and T. Sarkar. Delhi: Primus, 2015.

Palace, Wendy. *The British Empire and Tibet 1900–1922*. London: Routledge, 2004.

Panwar, Namrata. 'From Nationalism to Factionalism: Faultlines in the Naga Insurgency'. *Small Wars and Insurgencies* 28, no. 1 (2017): pp. 233–258.

Parratt, Saroj N. Arambam. *The Court Chronicle of the Kings of Manipur: The Cheitharon Kumpapa, Original Text, Translation and Notes, vol. 1, 33-1763 CE*. London and New York, NY: Routledge, 2005.

Parry, Nevill Edward. *The Lakhers*. London: Macmillan and Co. Ltd., 1932.

Peal, S. 'Note on the Old Burma Route over Patkai'. *Journal of the Asiatic Society of Bengal* XLVIII, no. II. (1879).

Pearn, Bertie Reginald. *Burma Background* (No. 1). London: Longmans Green & Co., 1943.

Pedersen, Morten Axel, and Bunkenborg, Mikkel. 'Roads that Separate: Sino-Mongolian Relations in the Inner Asian Desert". *Mobilities*, 7, no. 4 (2012): pp. 555-567.

Pemberton, Robert Boileau. *Report on the Eastern Frontier of British India*. Calcutta: Baptist Mission Press, 1835.

Phizo, Angami Zapu. *Plebiscite Speech* (1951), accessed 2 December 2021,http://www.neuenhofer.de/guenter/nagaland/phizo.html.

Pollak, Olliver B. 'The Origins of the Second Anglo-Burmese War (1852-53)'. *Modern Asian Studies* 12, no. 3 (1978): pp. 483–502.

Pollock, Sheldon. *The Language of the Gods in the World of Men: Sanskrit, Culture, and Power in Premodern India*. Berkeley and Los Angeles, CA: University of California Press, 2007.

Pommaret, Françoise. 'Ancient Trade Partners: Bhutan, Cooch Bihar and Assam' (17th to 19th centuries)' (2000), accessed 2 December 2021, http://himalaya.socanth.cam.ac.uk/collections/journals/jbs/pdf/JBS_02_01_02.pdf pp. 1–25.

Prasse-Freeman, Elliott. 'The Rohingya Crisis'. *Anthropology Today* 33, no. 6 (2017): pp. 1–2.

Prasse-Freeman, Elliott. 'Refusing Rohingya: Reformulating Ethnicity Amidst Blunt Biopolitics'. In *Current Anthropology* (forthcoming 2022).

Pratt, Mary Louise. 'Arts of the Contact Zone'. *Profession* (1991): pp. 33–40.

Pratt, Mary Louise. *Imperial Eyes: Travel Writing and Transculturation.* London and New York, NY: Routledge, 1992.

Proctor, Robert. 'Anti-Agate: The Great Diamond Hoax and the Semiprecious Stone Scam'. *Configurations* 9, no. 3 (2001): pp. 381–412.

Proctor, Robert. 'Agnotology: A Missing Term to Describe the Cultural Production of Ignorance (and Its Study)' In *Agnotology: The Making and Unmaking of Ignorance*, edited by R. Proctor and L. Shiebinger, pp. 1–33. Stanford, CA: Stanford University Press, 2008.

Public Relations Committee, Arakan Students Union. *Facebook*, accessed 31 July 2020, https://www.facebook.com/asunion.tcom/?tn-str=k*F.

Rahman, Abdur. *Denial and Deprivation: Indian Muslims after the Sachar Committee and Rangnath Mishra Commission Reports.* London: Taylor & Francis Group, 2018.

Rajesh, M. N. 'Tibet and Bengal, Trade Routes and Long Distance Exchange with Reference to Silver'. In *From Mountain Fastness To Coastal Kingdoms*, edited by J. Deyell and R. Mukherjee. Delhi: Manohar, 2019.

Rakhine Conflicts Investigation Commission. Press Release number 2, 5 November 2012, n.p.

Ray, Haraprasad. *Trade and Trade Routes Between India and China, C. 140 B.C.–A.D. 1500.* New Delhi: Progressive, 2003.

Reid, Adam Scott. *Chin-Lushai Land.* Calcutta: Thacker, Spink and Company, 1893.

Related Documents about Yunnan in Veritable Records of the Ming, Part 1. Kunming: People's Publishing House of Yunnan, 1959.

Reliefweb. 'Conflicts without borders', https://reliefweb.int/sites/reliefweb.int/files/resources/4307B727E9530039852576B10054EA19-map.pdf, accessed 23 February 2021.

Richards, John F. 'The Finances of the East India Company in India, c. 1766-1859'. Working Papers No. 153/11, London School of Economics, 2011.

Rienjang, Wannaporn, and Peter Stewart. *The Geography of Gandharan Art: Proceedings of the Second International Workshop of the Gandhāra Connections Project.* Oxford: Archaeopress, 2018.

Rippa, Alessandro. 'Centralizing Peripheries: The Belt and Road Initiative and its Role in the Development of the Chinese Borderlands'. *International Journal of Business Anthropology* 7, no. 1 (2019): pp. 1–22.

Rippa, Alessandro. 'Zomia 2.0: Branding Remoteness and Neoliberal Connectivity in the Golden Triangle Special Economic Zone, Laos'. *Social Anthropology* 27, no. 2 (2019): pp. 253–269.

Rippa, Alessandro. *Borderland Infrastructures: Trade, Development and Control in Western China.* Amsterdam: Amsterdam University Press, 2020.

Risor, Helene. 'Twenty Hanging Dolls and a Lynching: Defacing Dangerousness and Enacting Citizenship in El Alto, Bolivia'. *Public Culture* 22, no. 3 (2010): pp. 465–485.

Robinson, William. *A Descriptive Account of Assam: With a Sketch of the Local Geography and a Concise History of the Tea-plant of Assam.* London: Ostell and Lepag, 1841.

Rosanga, O. 'The Consolidation of Mizoram with Special Reference to the Disarmament of the Hills'. *Historical Journal Mizoram* VIII (2007 [1929]): pp. 98–108.

Roy, Kaushik. *Military Manpower, Armies and Warfare in South Asia.* London: Routledge, 2013.

Roy, Sandip. 'Don't Touch "our" Women: Dimapur Lynching was never about Justice' (2015), accessed 2 December 2021, https://www.firstpost.com/india/dont-touch-our-women-the-dimapur-lynching-was-never-about-justice-2143305.html.

Sacks, Milton. 'The Strategy of Communism in Southeast Asia'. *Pacific Affairs* 23, no. 3 (1950): pp. 227–247.

Sadan, Mandy. *Being and Becoming Kachin: Histories Beyond the State in the Borderworlds of Burma.* A British Academy Postdoctoral Fellowship Monograph. Oxford: Oxford University Press, 2013.

Sadan, Mandy. 'Introduction'. In *War and Peace in the Borderlands of Myanmar: The Kachin Ceasefire, 1994-2011,* edited by Mandy Sadan, pp. 1–28. Copenhagen: Nias Press, 2016.

Safonova, Tatiana, István Sántha, and Pavel Sulyandziga. 'Searching for Trust: Indigenous People in the Jade Business'. In *Trust and Mistrust in the Economies of the China-Russia Borderlands*, edited by Caroline Humphrey, pp. 205–229. Amsterdam: Amsterdam University Press, 2018.

Saikia, Arupjyoti. *Forests and Ecological History of Assam, 1826–2000.* New Delhi: Oxford University Press, 2011.

Saikia, Arupjyoti. *The Unquiet River: A Biography of the Brahmaputra.* New Delhi: Oxford University Press, 2019.

Sarar. 'Hopeless Cage'. *Facebook: 'The Art Garden Rohingya,* accessed 29 July 2020, https://m.facebook.com/HeartbeatsOfRohingya/photos/a.1720068411463546/1720068738130180/?type=3.

Sarar. 'Wish'. *Facebook: 'The Art Garden Rohingya',* accessed 28 July 2020, https://www.facebook.com/HeartbeatsOfRohingya/photos/a.1720068411463546/1790362767767443/?type=3.

Saxer, Martin. 'Pathways: A Concept, Field Site and Methodological Approach to Study Remoteness and Connectivity'. *Himalaya – The Journal of the Association for Nepal and Himalayan Studies* 36, no. 2 (2016): pp. 104–119.

Schober, Juliana. 'Belonging in a New Myanmar: Identity, Law, and Gender in the Anthropology of Contemporary Buddhism'. *Religion and Society: Advances in Research* 8 (2017): pp 158–172.

Schumann, Walter. *Gemstones of the World*. New York: Sterling, 1976.
Schwartzberg, Joseph E., ed. *A Historical Atlas of South Asia*. New York, NY: Oxford University Press, 1992.
Scott, James C. *The Art of Not Being Governed: An Anarchist History of Upland Southeast Asia*. New Haven, CN and London: Yale University Press, 2009.
Sen, Amartya. *Poverty and Famines: An Essay on Entitlement and Deprivation*. Oxford: Oxford University Press, 1981.
Sen, Tansen. 'The Travel Records of Chinese Pilgrims Faxian, Xuanzang, and Yijing: Sources for Cross-Cultural Encounters between Ancient China and Ancient India'. *Education about Asia* 11, no. 3 (2006): pp. 24–33.
Shaffer, Lynda. 'Southernization'. *Journal of World History* 5, no. 1 (1994): pp. 1–21.
Shakespear, John. *The Lushei Kuki Clans*. London: Macmillan, 1912.
Shakespear, Leslie Waterfield (Colonel). *History of the Assam Rifles*. London: Macmillan, 1929.
Shi, Shuo. *The Tibetan and Yi Corridor*. Chengdu: People's Publishing House of Sichuan, 2009.
Shilu, Gaozong. *Veritable Records of the Qing*. Gaozong Shilu, scroll 804, Vol.18. Beijing: Zhonghua Book Company, 1985.
Shimray, R. R. *Origin and Culture of Nagas*. New Delhi: Somsok Publications, 1985.
Shneiderman, Sara. 'Are the Central Himalayas in *Zomia*? Some Scholarly and Political Considerations Across Time and Space'. *Journal of Global History* 5, no. 2 (2010): pp. 289–312.
Sigley, Gary. 'Tea and China's Rise: Tea, Nationalism and Culture in the 21st Century'. *International Communication of Chinese Culture* 2, no. 3 (2015): pp. 319–341.
Simmel, George. 'The Stranger'. In Kurt Wolff (transl.), *The Sociology of Georg Simmel*, pp. 402–408. New York, NY: Free Press, 1950.
Singh, Amerjeet. *A Study on Illegal Immigration into Northeast India: The Case of Nagaland*. IDSA Occasional Paper no. 8 (2009), accessed 2 December 2021, https://idsa.in/system/files/OccasionalPaper8_NagalandIllegalImmigrationl.pdf.
Singh, B. P. 'North-East India: Demography, Culture and Identity Crisis'. *Modern Asian Studies* 21, no. 2 (1987): pp. 257–282.
Sinha, Surajit, ed. *Tribal Polities and State Systems in Pre-Colonial East and North-eastern India*. Calcutta: KP Bagchi and Co., 1987.
Siraj, Nasreen, and Ellen Bal. '"Hunger has Brought us into this Jungle": Understanding Mobility and Immobility of Bengali Immigrants in the Chittagong Hills of Bangladeshi'. *Social Identities* 23, no. 4 (2017): pp. 396–412.
Skinner, G. William. 'Marketing and Social Structure in Rural China: Part I'. *The Journal of Asian Studies* 24, no. 1 (Nov. 1964.): pp. 3–43.
Slack, Paul. 'Responses to Plague in Early Modern Europe: The Implications of Public Health'. *Social Research* 55, no. 3 (1988): pp. 433–453.

Smith, Martin. 'Reflections on the Kachin Ceasefire: A Cycle of Hope and Disappointment'. In *War and Peace in the Borderlands of Myanmar: The Kachin Ceasefire, 1994-2011*, edited by Mandy Sadan, pp. 57–91. Copenhagen: Nias Press, 2016.

Smyer Yü, Dan. 'Introduction: Trans-Himalayas as Multistate Margins'. In *Trans-Himalayan Boundaries: Livelihoods, Territorialities, Modernities*, edited by J. Michaud and D. Smyer Yü, pp. 11–42. Amsterdam, Amsterdam University Press, 2017.

Smyer Yü, Dan. 'Trans-Himalayas as Multistate Margins'. In *Trans-Himalayan Borderlands: Livelihoods, Territorialities, Modernities*, edited by Dan Smyer Yü and Jean Michaud, pp. 8–41. Amsterdam: Amsterdam University Press, 2017.

Smyer Yü, Dan, and Karin Dean, eds. *Yunnan–Burma–Bengal Corridor Geographies: Protean Edging of Habitats and Empires*. New Delhi: Routledge, 2021.

Sneddon, Christopher. *Understanding Kashmir and Kashmiris*. London: Hurst and Company, 2015.

Stein, Burton. 'Integration of the Agrarian System of South India'. In *Land Control and Social Structure in Indian History*, edited by Robert Eric Frykenberg, pp.175–216. Madison, WI: University of Wisconsin Press, 1969.

Stevenson, Henry Noel Cochran. *The Hill Peoples of Burma* (No. 6). London: Longmans, Green & co., Ltd., 1945.

Steyn, Peter. *Zapuphizo: The Voice of the Nagas*. London: Routledge, 2002.

Stonor, C. R., and Edgar. Anderson. 'Maize among the Hill Peoples of Assam'. *Annals of the Missouri Botanical Garden* 36, no. 3 (1949): pp. 355–404.

'Students Protest Rape in Kohima'. *The Telegraph*, 4 March 2015.

Subrahmanyam, Sanjay, and Christopher A. Bayly. 'Portfolio capitalists and the *political* economy of early modern India'. *The Indian Economic and Social History Review* 25, no. 4 (1988): pp. 401–424.

Sullivan, Joanna. 'Redefining the Novel in Africa'. *Research in African Literatures* 37, no. 4 (2006).

Sun, Laichen. 'From Baoshi to Feicui: Qing-Burmese Gem Trade, c. 1644-1800'. In *Chinese Circulations: Capital, Commodities and Networks in Southeast Asia*, edited by E. Tagliacozzo and W-C. Chang, pp. 203–221. Durham, NC and London: Duke University Press, 2011.

Swu, Scato. *Hails and Blames: A Brief Account of Naga Independence Struggle*. Dimapur: Heritage Publications, 2013.

Taussig, Michael. *Defacement: Public Secrecy and the Labor of the Negative*. Stanford, CA: University Press, 1999.

Than, Tharaphi. 'Women in Modern Burma'. *Southeast Asian Studies* 6, no. 1 (April 2017): pp. 167-170.

Thapar, Romila. *Somanatha: The Many Voices of a History*. London: Verso, 2005.

Thaung, Ma. 'British Interest in Trans-Burma Trade Routes to China, 1826-1876'. London: University of London, 1954.

'The Medical and Physical Society of Calcutta'. *Asiatic Journal and Monthly Register for British India and Its Dependencies* 25 (January–June 1828): pp. 225–226.

Thein, Pe. *My Autobiography: Good and Bad*. Yangon: Electricity Publishing, 2009.

Thohe Pou, R. B. 'The Myths of Naga Origin', published 2007, accessed 2 December 2021, https://thohepou.wordpress.com/2007/08/17/the-myths-of-naga-origin/.

Tinyi, Venusa. 'The Headhunting Culture of the Nagas: Reinterpreting the Self'. In *Nagas in the 21st Century*, edited by J. J. P. Wouters and M. Heneise, pp. 117–132. Kohima and Edinburgh: The Highlander Books, 2017.

Tong, Shaoyu and Chen, Yongsen. *The Study of Bazi in Yunnan*. Kunming: Yunnan University Press, 2007.

Tsing, Anna Lowenhaupt. *Friction: An Ethnography of Global Connection*. Princeton, NJ: Princeton University Press, 2005.

Uddin, Nasir, and Nasreen Chowdhory, eds. *Deterritorialised Identity and Transborder Movement in South Asia*. Singapore: Springer, 2019.

United Nations. 'Global issues, refugees', accessed 4 August 2021, https://www.un.org/en/global-issues/refugees.

Uribe, Maria. V. 'Dismembering and Expelling: Semantics of Political Terror in Colombia'. *Public Culture* 16, no. (2004): pp. 79–96.

Van Schendel, Willem. 'Geographies of Knowing, Geographies of Ignorance: Jumping Scale in Southeast Asia'. *Environment and Planning D: Society and Space* 20 (2002): pp. 647–668.

Van Schendel, Willem. *The Bengal Borderland: Beyond State and Nation in South Asia*. London: Anthem Press, 2004.

Van Schendel, Willem. 'The Dangers of Belonging: Tribes, Indigenous Peoples and Homelands in South Asia'. In *The Politics of Belonging in India: Becoming Adivasi*, edited by D. J. Rycroft and S. Dasgupta, pp. 19–43. London and New York, NY: Routledge, 2011.

Van Schendel, Willem. *Embedding agricultural commodities: using historical evidence, 1840s-1940s*. London: Routledge, 2017.

Van Schendel, Willem. *A History of Bangladesh*. Cambridge: Cambridge University Press, 2020 [2009].

Van Spengen, Wim. *Tibetan Border Worlds: A Geohistorical Analysis of Trade and Traders*, London, Routledge, 2000.

Vann, Elizabeth. 'Production Matters: Consumerism and Global Capitalism in Vietnam'. In *Research in Economic Anthropology: Anthropological Perspectives on Economic Development and Integration*, vol. 22, edited by N. Dannhaeuser and C. Werner, pp. 225–257. London: Emerald Group Publishing, 2003.

Verardi, Giovanni. *Hardships and Downfall of Buddhism in India*. New Delhi: Manohar, 2011.

Vogel, Hans Ulrich, and Sabine Hieronymus. 'Cowry Trade and Its Role in the Economy of Yunnan: From the Ninth to the Mid-Seventeenth Century'. *Journal of the Economic and Social History of the Orient* 36, no. 3 (1993): pp. 211–252.

Waddell, Laurence Austine. *The Buddhism of Tibet or Lamaism*. London: W.H. Allen & Co., 1895.

Wade, Francis. *Myanmar's Enemy Within: Buddhist Violence and the Making of a Muslim 'Other'*, 2nd ed. London: Zed Books, 2017.

Walker, J. 'Jade: A Special Gemstone'. In *Jade*, edited by R. Keverne. New York: Springer.

Wallich. Nathaniel. 'Notes on the Drug called Mishmi Teeta and Pucha Pat'. *Madras Journal of Literature and Science* 5 (1837): p. 347.

Wallich. Nathaniel. 'Medico-Botanical Notices'. *American Journal of Pharmacy* 9 (1838): p. 196.

Walsh, Andrew. 'In the Wake of Things: Speculating in and about Sapphires in Northern Madagascar'. *American Anthropologist* 106, no. 2 (2004): pp. 225–237.

Wang, Kangguo. 'Draft of Zhaozhou Gazetteer'. In *Collection of Historical Inscriptions in Fengyi*, edited by Ma Cunzhao. Hong Kong: South China Research Center, 2013.

Wardle, Thomas. *Empire of India: Special Catalogue of Exhibits by the Government of India and Private Exhibitors, Colonial and Indian Exhibition, 1886*. London: William Clowes and Sons, 1886.

Waring, Edward John. *Pharmacopoeia of India*. London: W.H. Allen, 1868.

Waterfield, Henry. *Memorandum on the Census of British India 1871-72*. London: Eyre and Spottiswoode, 1875.

Watt, George. *A Dictionary of the Economic Products of India*, vol. 2. Calcutta: Superintendent of Government Printing, 1889.

Way, R. A. *Assam-Burma Connection Railway Surveys, Hukong Valley Route: Report and Approximate Estimate, 1896*. Calcutta: Superintendent of Government Printing, 1896.

Webb, Matthew J. *Separatist Violence in South Asia: A Comparative Study*. London: Taylor and Francis, 2016.

White, Richard. *The Middle Ground: Indians, Empires, and Republics in the Great Lakes Region, 1650-1815*. Cambridge Studies in North American Indian History. Cambridge and New York, NY: Cambridge University Press, 1991.

'Who are the Rohingya? The More than One Million Rohingya Muslims are Described as the "World's Most Persecuted Minority"'. *Aljazeera*, 18 April 2018.

Williamson, Noël. 'The Lohit-Brahmaputra between Assam and South-Eastern Tibet, November 1907 to January 1908'. *The Geographical Journal* 34, no. 4 (1909): p. 364.

Willoughby, M. E. *Report on the Mules and Ponies of the Province of Yünnan (West China)*. Calcutta: Office of the Superintendent of Government Printing, India, 1904.

Wimmer, Andreas, and Nina Glick Schiller. 'Methodological Nationalism and Beyond: Nation-State Building, Migration and the Social Sciences'. *Global Networks* 2, no. (2002): pp. 301–334.

Wink, André. *Al-Hind, The Making of the Indo-Islamic World: Slave Kings and the Islamic Conquest, 11th-13th Centuries*, vol. 2. Boston: Brill, 2002.

Wood, Mary. *Lynching and Spectacle: Witnessing Racial Violence in America, 1890-1940*. Chapel Hill, NC: The University of North Carolina Press, 2009.

Wood, Thomas. *Map Showing Route from Gwahatty to Kolltah Koosy on the Borders of Bootan*, no. 1, in Private Papers of Francis Buchanan-Hamilton, IOR Mss Eur D. 97, Asia and Africa Collection, British Library, 1793.

Woods, Kevin. 'Ceasefire Capitalism: Military-Private Partnerships, Resource Concessions and Military State Building in the Burma-China Borderlands'. *Journal of Peasant Studies* 38, no. 4 (2011): pp. 747–770.

Woods, Kevin. 'Conflict Timber along the China-Burma Border: Connecting the Global Timber Consumer with Violent Extraction Sites'. In *Chinese Circulations: Capital, Commodities, and Networks in Southeast Asia*, edited by E. Tagliacozzo and W-C. Chang, pp. 480–507. Durham, NC and London: Duke University Press, 2011.

Woods, Kevin. 'The Commercialisation of Counterinsurgency: Battlefield Enemies, Business Bedfellows in Kachin State, Burma'. In *War and Peace in the Borderlands of Myanmar: The Kachin Ceasefire, 1994-2011*, edited by Mandy Sadan, pp. 114–145. Copenhagen: Nias Press, 2016.

Woodthorpe, Robert Gosset. *The Lushai Expedition, 1871-1872*. London: Hurst and Blackett, 1873.

Wouters, Jelle J. P. 'Keeping the Hill Tribes at Bay: A Critique from India's Northeast of James C. Scott's Paradigm of State Evasion'. *European Bulletin of Himalayan Research* 39 (2011): pp. 41–65.

Wouters, Jelle J. P., and T. B. Subba. 'The "Indian Face," India's Northeast, and "the Idea of India"'. *Asian Anthropology* 12, no. 3 (2013): pp. 126–140.

Wouters, Jelle J. P. 'Sovereignty, Integration or Bifurcation? Troubled Histories, Contentious Territories, and the Political Horizons of the Long Lingering Naga Movement'. *Studies in History* 32, no. 1 (2016): pp. 97–116.

Wouters, Jelle J. P. *In the Shadows of Naga Insurgency: Tribes, State, and Violence in Northeast India*. New Delhi: Oxford University Press, 2018.

Wouters, Jelle J. P. 'Difficult Decolonization: Debates, Divisions, and Deaths within the Naga Uprising, 1944-1963'. *Journal of North-East India Studies* 9, no. 1 (2019): pp. 1–28.

Wouters, Jelle, J. P. 'Neoliberal Capitalism and Ethno-Territoriality in Highland Northeast India: Resource-Extraction, Capitalist Desires and Ethnic Closure'. *Geopolitics* (2020). Doi: 10.1080/14650045.2020.1812581.

Yang, Bin. 'Horses, Silver, and Cowries: Yunnan in Global Perspective'. *Journal of World History* 15, no. 3 (September 2004): pp. 281–332.

Yang, Bin. *Between Wind and Clouds: The Making of Yunnan (Second Century BCE to Twentieth Century CE)*. New York, NY: Columbia University Press, 2009.

Yang, Bin. 'The Bay of Bengal Connections to Yunnan'. In *Pelagic Passageways: The Northern Bay of Bengal Before Colonialism*, edited by Rila Mukherjee, pp. 317–342. Delhi: Primus Books, 2011.

Yang, Bin. 'The Rise and Fall of Cowrie Shells: The Asian Story'. *Journal of World History* (2011): pp. 1-26.

Yang, Bin. 'The Laws of King Mengrai, Cowrie Shells and Intra-Asian Interactions'. In *From Mountain Fastness to Coastal Kingdoms*, edited by J. Deyell and R. Mukherjee. Delhi: Manohar, 2019.

Yang, Jiajing. *The Fengyi Gazetteer*. Kunming: Yunnan University Press, 1996.

Yang, Shen, and Wei Hu. *The Unofficial History of Nanzhao (1880)*. Taipei: Chengwen Publising House, 1968.

You, Zhong. *Collected Works of You Zhong*, vol. 2. Kunming: Yunnan University Press, 2009.

Young, E. C. 'A Journey from Yün-Nan to Assam'. *The Geographical Journal* 30, no. 2 (1907): pp. 152–180.

Yu, Ming. *Chinese Jade*. Cambridge: Cambridge University Press, 2009.

Yule, Henry. 'Notes on the Iron of the Kasia Hills, for the Museum of Economic Geology'. *Journal of the Asiatic Society of Bengal* 11, no. 2 (1842): pp. 853–857.

Yule, Henry. 'On the Geography of Burma and its Tributary States, in Illustration of a New Map of those Regions'. *The Journal of the Royal Geographical Society of London* 27 (1857): pp. 54–108.

Zhang, Jinghong. *Puer Tea: Ancient Caravans and Urban Chic*. Seattle, WA and London: University of Washington Press, 2014.

Zhao, Min. 'Salt, grain and the change of deities in early Ming western Yunnan'. In *The Transformation of Yunnan in Ming China: From the Dali Kingdom to Imperial Province*, edited by C. Daniels and J. Ma, pp. 19–42. London: Routledge, 2019.

Zhuang, Cheng, ed. *The Wanli Gazetteer of Zhaozhou Subprefecture* (1587). Dali: Reprinted by the Prefectural Bureau of Cultural Affairs, 1983.

Žižek, Slavoj. *Did Somebody Say Totalitarianism? Five Interventions in the (Mis)use of a Nation*. London: Verso, 2005.

Zolberg, Aristide R. 'The Formation of New States as a Refugee-Generating Process' in 'The Global Refugee Problem: U. S. and World Response', special issue, *The Annals of the American Academy of Political and Social Science* 467 (May 1983): pp. 24–38.

Index

adjacent identities 238, 249
agnotology 207, 223
Ahom *see* Assam, Tai-Ahom in
All under Heaven cosmology 53-75
Appadurai, Arjun 173, 183, 187
Arakan *see* Rakhine
archives 22-23, 60, 79, 106-107, 143, 153
armed conflict 12, 31, 92, 169, 176, 215, 220
Arunachal Pradesh 107, 109, 112, 116-118, 125-126, 186, 196-197
Assam 16-17, 43, 45, 82, 106, 116-118, 123-125, 128-129, 137-163, 178-179, 181, 186, 191, 208
  Assamese 21, 23, 112, 119, 130, 141-3, 156, 172, 179
  Tai-Ahom in 109-112, 114, 119, 122, 124-125, 172

Bangladesh 12, 16-17, 21, 33,43, 46, 167-202, 236
Barak 31, 43, 109, 129
barbarian 54, 69-70, 72, 83, 94
basin society 53-75
Bay of Bengal 13, 16, 111, 154; *see also* Indian Ocean
bazi 53-75
Beijing (Peking) 41, 70, 97, 209, 211
Belt and Road Initiative (BRI) 31, 210, 216, 225-126
Bengal 30, 33-40, 84, 109-114, 121-125, 127, 129-131, 141-142, 146, 149, 151-154, 172, 178, 243
  cowries in 31, 112, 116-117, 131, 143,
  Bengali 20, 43, 112, 172, 216, 236-237, 240, 247-248
Bhutan 12, 109-110, 130, 141, 143, 150
Black Lives Matter 29-30, 40, 46
Blackburn, Stuart 107
border 12, 22, 31, 43, 77-103, 107-108, 117, 125-126, 129, 154-157, 172, 205, 209-214, 231-232, 236, 250
  mental border 21, 231
  nation-state border 11-12, 45-46, 82, 108, 157, 178-179, 219, 223-225
borderland 11-13, 16, 45, 54, 56, 62, 70-71, 77, 80, 84, 99, 105, 205, 207, 209-211, 219-217, 221, 225, 234
borderworld 22, 234
Brahmaputra 31, 39, 83, 109-110, 129, 140-141, 154
Buddhist/ism 20, 34-40, 45-46, 60, 66, 68, 111, 125-126, 140, 204-205, 216, 221-222, 233, 243-244, 249
buffer zone 80
Burma 17, 21, 34, 44, 46, 53-75, 105-137, 140, 143-144, 146-147, 176-177, 196, 203-229; *see also* Myanmar
  British Burma 31, 40-41, 77-103, 151, 154, 156

Cachar 83-84, 110, 114, 120-121, 128-129
Calcutta *see* Kolkata
caravan 18-19, 38, 58-59, 70, 79-80, 85-87, 89-91, 93, 97-98, 116, 207, 209-210
Carey, Bertram Sausmarez 114, 116, 120-122
Carey, Felix 83
Carey, M. 206, 225
Cederlöf, Gunnel 17-18, 20, 22-23, 55-56, 224
Charney, Michael 127, 243
Chin, Chin State 15, 109, 120-123, 236
China 11-25, 30-35, 40-42, 44-45, 53-75, 77-103, 109, 112, 119, 130, 137, 140-146, 150, 154-157, 203-229, 248
Chindwin 31, 109, 125, 127
Chittagong 117, 120-121, 128, 130
Chittagong Hill Tracts 40, 109-110, 120-121, 126, 172, 233
Christian, Christianity 20-21, 35, 40, 126-127, 171, 181, 192
circulation 56, 105-136
Confucian 40, 54, 57-58, 66, 204, 222
contact zone 81, 216
corridor *see* India-China Corridor
Covid-19 29-30, 40
cowry 31, 105, 108, 111-112, 116-118, 123-124, 131, 143

de Certeau, Michel 30
digital 21, 235, 238
digital participant observation 235
Dimapur 167-171, 177-182, 186, 188-191
distance-demolishing technologies 17
Douglas, Mary 173, 188
drug 12, 16, 22, 31, 43, 129, 143, 145-6, 204, 207-210

East India Company (British) 42-43, 77-103, 137, 139-147, 149, 152, 208
eastern Himalayas 11-14, 109, 137-157, 232
EIC *see* East India Company (British)
empire 30-31, 33, 35-36, 39-40, 42-44, 46, 57, 60, 64, 71, 78, 80, 84, 91, 119, 137, 139, 148, 155-157

Fang Guoyu 57
Fengshui 58, 68-70
flow 11-25, 54-59, 70, 77-79, 85-87, 89, 96-100, 105, 124-127, 129, 137, 141-57, 167, 170-176, 186, 205-208, 212, 216, 218; *see also* mobility
friction 14-19, 23, 29, 31, 100, 167, 170-171, 176, 203-207, 211, 215, 217-218, 224-225
frontier 32-33, 37, 39-40, 42-43, 46, 56-61, 63-64, 70, 79-83, 137-163
  frontier society 57
  frontier zone 80
Fur, Gunlög 81

Giersch, C. Patterson 56, 80-81
glacier 17
Golden Triangle 22
gun 12, 16, 31, 119-123, 176, 189

Hechter, Michael 57
Herman, John 57
Hill, Maxwell 56
Hindu, Hinduism 20, 35-36, 38-40, 42, 45-47, 124-126, 183, 195, 243
Hostetler, Laura 57
Hui 60, 86, 90, 116
Hukawng 117-119

ID card 234, 236-237, 247-248
ignorance, production of 19-23, 205-206, 221, 223-225
India 12-14, 16-23, 31, 33-34, 36, 207, 216, 224-225, 247
  British India 40, 42-45, 47. 82, 85, 91-92
  Northeast India 105-163, 167-202, 238-239
  Republic of India 12-14, 46-47
India-China Corridor 13-15, 16-17, 18-20, 22-23, 31-33, 45, 56-57, 170, 210, 224-225, 232, 238, 240, 250
Indian Ocean 17, 29-30, 32-33, 39, 42, 80, 111
Indo-Burma Range 13, 16-17, 105, 109
internal colonialism 57
internal Other 187, 194-195
Irrawaddy, Ayeyarwadi 31, 40, 43, 59, 80, 87, 99, 109, 119, 130-131, 154
Islam 35, 38-39, 125, 180, 235, 244

jade 12, 17, 21, 203-229
Jhala, Angma 110
Jinghpaw *see* Kachin

Kachin 16-17, 94, 118-119, 126, 128, 147-148, 203-229, 236, 241
Karnafuli 129
Kikon, Dolly 178, 191
Kolkata 43-44, 91, 117-118, 120, 145, 148, 151
Kuki 120-121, 123, 126
Kunming 59, 62, 86, 208

Laos 71, 78, 83, 89, 196
Leach, Edmund 58, 207
Ledo Road 31
Lee Guo Lun 85, 94, 98
Ludden, David 17, 23, 117, 231-232, 236
Lushai hills *see* Mizoram

Ma, Jianxiong 17-18, 20, 22-23, 80-82, 85, 116
Manipur 40, 82, 105-106, 109-110, 112-114, 116, 118, 123, 125-129, 156, 186, 196-197
May, Htoo 14, 17, 21-23, 216
Meghalaya 43, 109, 142, 197
Meghna 32, 39-40, 43
Mekong 53

mental border 21
middle ground 56, 80-82
Middleton, Townsend 173
Mizoram 14, 17, 20, 105-136
mobility 29-50, 139, 175-177, 210, 221, 231-53; *see also* flow
Møller, Henrik Kloppenborg 17-18, 20-21, 23
multistate margin 22, 140
Muslim *see* Islam
Myanmar (Burma) 12, 14, 16-17, 21, 33, 41, 45, 109, 114, 118, 169, 176, 186, 196, 203-217, 220-6, 231-253; *see also* Burma
Myitkyina 207, 216, 218-20

Naf 40
Nagaland, Naga 14, 16-18, 21, 40, 107, 109, 112, 116-118, 122, 128, 167-202, 216
Nagamese 20
nation-state 12-13, 16, 19, 41, 57, 80, 82, 99, 155-6, 174, 183, 186, 210, 222-223
non-state space 80

opacity 18, 203, 206-207, 211, 214-215, 217, 221, 224-226

Pachuau, Joy L.K. 14, 17, 23, 224
pandemic 30, 42, 249
Panthay Rebellion 77-78, 83, 85-86, 89-100, 116, 208
pathway 15-16, 18-20, 78, 89, 137
Patkai 114, 128, 137, 151, 153-154, 156, 158, 161
Pedersen, M.A. 206, 225
Pegu (Bago) 98, 117
Peking *see* Beijing
Pratt, Mary Louise 81, 225

Rakhine (Arakan) 14, 16-17, 20, 32, 34, 39-40, 43, 46, 109-110, 117, 120-121, 125, 128, 130-131, 216, 231-253
Rangoon *see* Yangon
Rohingya 21, 46, 216, 231-253
Ruili 210, 212, 214, 219, 248

Sadan, Mandy 93, 204, 208, 211, 241, 250
Saikia, Arupjyoti 17, 23
Salween 53
Sanskrit cosmopolis 37
SARS-Cov-19 *see* Covid-19
sawbwa 55, 79-80, 84-85, 91-100
Saxer, Martin 89
Scott, David 110, 144
Scott, James C. 55, 81, 196
Scott, James George 79
Second World War 18, 31, 137, 139, 156-157, 225
Shan 91-97, 109, 116, 119, 123, 128, 209-210, 212, 219, 236
Shan-Dai 16-18, 55, 61, 77-93
Shanghai 203, 205-206, 209, 221, 223
Sichuan 18, 89, 109, 207

Sikkim 109-110
Silk Road 32, 34-36, 38, 41, 46, 80, 207
Singpho *see* Kachin
Sittwe 14, 231-253
Skinner, William 59
Sladen, Edward B. 77-78, 80, 83-87, 89-100.
Southeast Asian Massif 22
spatial history 11, 23, 29-50
Sylhet 32, 39, 43, 45, 114, 117, 128-129, 142-143

Tani 107
Taoist 119, 204, 209
tax 18, 21, 43, 54, 58, 62-63, 68, 141, 146, 172, 178, 186, 203, 205-206, 208, 211-212, 214-215, 218-220, 224-225
Tengchong (Momein) 77, 80, 212, 220
territoriality 21, 29, 41-42, 175
Thailand (Siam) 12, 40, 45, 56, 71, 82, 89, 111, 118, 119, 209-210, 212, 214, 219
Than, Tharaphi 14, 17, 21-23, 216
Tibet 18, 31-33, 36, 38, 42-43, 54, 59, 68, 70, 81, 89, 107, 110-112, 124-126, 131, 141-143, 145, 149-151, 154-156, 205, 207
time-space compression 17-18
Trans-Himalayan spaces 11-23
tribe 14, 21, 40, 106, 109, 114, 116-117, 120-124, 128, 130-131, 144, 147-148, 150, 169-175, 177-178, 181-183, 185-187, 195, 197, 232, 238-239
tribute 53, 96, 100, 141-142, 208, 211, 225
Tripura 105, 109, 125, 128, 186
Tsing, Anna 14, 205, 224
Tuck, Henry Newman 114, 116, 120-122

Van Schendel, Willem 56, 107, 224, 238, 240, 245
verticality 17
Vietnam 71, 119

Westphalian 12
White, Richard 81
Wouters, Jelle J.P. 17, 21, 23, 216

Yang, Bin 108, 111, 117
Yangon (Rangoon) 44, 78, 84, 91, 98-99, 156, 210, 212, 214, 235, 248
Yangtze 53, 59
You Zhong 57
Yunnan 13, 16-18, 31-32, 40-43, 53-75, 77-103, 109, 111-112, 116-119, 124, 130-131, 155, 203-229

Zomia 22, 107

For Product Safety Concerns and Information please contact our EU representative GPSR@taylorandfrancis.com Taylor & Francis Verlag GmbH, Kaufingerstraße 24, 80331 München, Germany

Batch number: 10725973

Printed by Printforce, the Netherlands